Chemistry of Coal: Knowing the Basics

CHEMISTRY OF COAL: KNOWING THE BASICS

Edited by:

Dr. Kailash Malode

ARCLER PRESS

www.arclerpress.com

Chemistry of Coal: Knowing the Basics

Dr. Kailash Malode

Arcler Press
224 Shoreacres Road
Burlington, ON L7L 2H2
Canada
www.arclerpress.com
Email: orders@arclereducation.com

ISBN: 978-1-77469-822-8 (Hardcover)

ABOUT THE EDITOR

Dr. Kailash Malode (1980) is presently serving as Assistant Professor, Department of Soil Science And Agriculture Chemistry, college of Agriculture, Parul University, Vadodara, Gujarat. He obtained his B. Sc.(Ag.) in 2006 from Marathwada Agriculture University, Parbhani. M. Sc. (Ag.)2009 in Dr. Panjabrao Deshmukh Krishi Vidyapeeth, Akola and Ph. D. Soil Science And Agriculture Chemistry in 2014 from Vasantrao Naik Marathwada Krishi Vidhyapeeth, Parbhani. He started his career as Assistant professor in 2014. His field of specialization is soil fertility, soil carbon sequestration and organic carbon management. He has edited 06 books at international level, published 19 research papers in National and international journals of reputed. He has participated in many National, State Seminars and symposiums. He has written 02 practical manuals for the undergraduate course for the benefit of students, besides several radio talks and on farmers on a farmer training programme. Dr. Kailash has also written 26 popular articles in Shati Bharti, Shati pragati and other reputed magazines and 22 audio talks. He had the Best writer award in 2015 for Krishi dooth popular Magazines of Maharashtra, Young Scientists Award 2021 by Astha foundation and Birsa Agriculture University, Jharkhand. Dr. Kailash got the Best scientist award in soil science 2023 by Rajmata Vijayaraje Sindhya Krishi Vishva Vidhyalaya, Gwalior, and also best poster presentation of same.

TABLE OF CONTENTS

LIST OF FIGURES

LIST OF TABLES

LIST OF ABBREVIATIONS

\$/MWh	Dollars per Megawatt-Hour
AMD	Acid Mine Drainage
ASEAN	Association of Southeast Asian Nations
ASTM	American Society for Testing and Materials
BS	British Standards Institution
CCUS	Carbon Capture, Utilization, and Storage
CIAB	Coal Industry Advisory Board
CO_2	Carbon Dioxide
CO_2H	Carboxyl Chains
CRI	Certainty of Response Index
CSR	Coke Energy After Reaction
CTL	Coal-To-Liquids
CTPV	Coal-Tar Pitch Volatiles
CV	Calorie Value
CWP	Coal Workers Pneumoconiosis
DTA	Differential Thermal Analysis
DWT	Dead Weight Tonnage
GCCSI	Global CCS Institute
GCV	Gross Calorific Value
HELE	High-Efficiency, Low-Emissions
HGI	Hardgrove Grindability Index
HHV	Higher Heating Value
IEA	International Energy Agency
IGCC	Integrated Combustion Combined Cycle
INDCs	Intended Nationally Determined Contributions
ISO	International Standard Organization
LCOE	Levelized Cost of Electricity

LCTPI	Low Carbon Dioxide Technology Partnerships Initiative
LHV	Lower Calorific Value
NCV	Net Calorific Value
NOx	Nitrogen Oxide
OCGT	Open Cycle Gas Turbines
OECD	Organization for Economic Co-Operation and Development
ROM	Run-of-Mine
SC	Supercritical Carbon Dioxide
THF	Tetrahydrofuran
UNFCCC	United Nations Framework Convention on Climate Change
USC	Ultra-Supercritical Coal
XANES	X-Ray Absorption Near Edge Structure

PREFACE

Coal is a fossil fuel formed from the remains of plants that lived millions of years ago. The chemical composition of coal varies depending on the type of plant material that it is made from. Still, it typically contains high carbon, hydrogen, oxygen, nitrogen, and sulfur levels.

The carbon in coal is what gives it its energy value. When coal is burned, the carbon reacts with oxygen to form carbon dioxide and water vapour. This releases heat energy, which can be used to generate electricity or power machinery.

The hydrogen in coal also contributes to its energy value. When coal is burned, the hydrogen reacts with oxygen to form water vapour. This releases heat energy, but not as much as the carbon does. The oxygen in coal is not as important for its energy value, but it does play a role in the combustion process. The oxygen helps to fuel the fire and to prevent the coal from smoldering.

The nitrogen and sulfur in coal are considered to be impurities. They do not contribute to the energy value of coal, and they can make it more difficult to burn. Nitrogen can form harmful pollutants when burned, and sulfur can produce sulfur dioxide, a major contributor to acid rain. The chemistry of coal is complex, but the basic principles are relatively simple. By understanding the chemistry of coal, we can better understand how it burns and how it can be used to generate energy.

This book comprises of eight-chapters; the first chapter provides an introduction of coal with emphasis on the fundamentals of coal such as coal formation, coal classification, the discovery of coal and coal exploration. The second chapter deals with the precursors of coal and also discusses the heteroatoms in coal.

The third chapter provides insight into the preparation of coal along with its size reduction for the purpose of transportation. This chapter also discusses the storage mechanism of coal. The fourth chapter discusses the analysis techniques of coal.

The fifth chapter describes the different physical and electrical properties of coal as knowledge of these properties is also an important aspect of coal characterization and behavior. The sixth chapter deals with the thermal and mechanical properties of coal. The seventh chapter discusses the chemical reactivity of coal as an understanding of the chemical behavior of coal is an essential part of projecting the successful use of coal and the eighth chapter provides an insight about the impacts of coal utilization on health and safety.

The chemistry of coal is intended as a comprehensive resource for students, researchers, and professionals in the field of chemistry, energy, and materials science. It provides a solid foundation for understanding the complex nature of coal and its role in our world, emphasizing the importance of sustainable utilization.

CHAPTER

1

Introduction to the Chemistry of Coal

Contents

INTRODUCTION

The term "coal" is employed in a generic sense throughout the book to encompass all coal varieties. The material is a sedimentary rock that is organic and for biochemical derivation, characterized by its black hue, combustibility, and occurrence in rock layers, such as coal beds as well as coal seams (Orem & Finkelman, 2003). Coal mainly consists of carbon, along with various amounts of hydrogen, oxygen, nitrogen, as well as sulfur. Coal is found in layers known as seams or just strata. The USA possesses a considerable reserve for coal, comprising almost 30% of the global reserves, as well as a remained coal resource that exceeds 1,600 billion tons (1,600 × 109 tons). According to Höök & Aleklett (2009) as well as the sources they reference, the US is ranked as the second-biggest coal producer globally, with China holding the top position. Furthermore, the US generates over two times the amount of coal annually compared to India, which is ranked third in terms of coal production. Sequences of considerable thickness, comprising coal, sandstone, and shale rock, along with limestone storage, are frequently discovered in close proximity (Scott, 2002).

The Mississippian along with Pennsylvanian eras are acknowledged in the US owing to the notable rock successions discovered in those nations, specifically Mississippi along with Pennsylvania. The Cretaceous, Triassic, as well as Jurassic times, are recognized as significant epochs for coal deposition. The productivity of rocks tends to decrease at earlier ages, while younger deposits exhibit a prevalence of lignite along with peat. It is generally observed that coal reserves of higher positions are found in older strata. The economic worth of coal over a particular end use depends on its grade (Davis & Gerlach, 2018). The classification of coal pertains to the quantity of minerals contained within the coal and serves as an indicator of the coal's overall quality. Coal grading is determined by several factors, including the sulfur material, debris fusion temperatures (which refer to the temperatures when the residue melts as well as fuses), along with the number of minor elements present within the coal. While there is a lack of formal categories based on coal levels, the quality of coal remains a crucial factor for coal users. Coal is classified as a metamorphic rock owing to its formation through the application of extreme temperatures and pressures upon organic sediments like peat (Yu et al., 2007). The discourse supporting the categorization of coal to be sedimentary rock focuses on the fact that the majority of sedimentary rocks go through high temperatures and high pressures, and coal shares similar characteristics with common sedimentary rocks. Additionally, the formation process of coal typically places low-grade coal within the sedimentary categorization system. In contrast, anthracite is subjected to elevated levels of pressure as well as heat alongside is commonly linked to low-grade metamorphic formations like slate as well as quartzite (Schopf, 1948). Coal that has undergone subduction has the potential to transform into graphite within igneous rocks or just carbonate-rich rocks, such as carbonatites. Coal seems a fossil fuel that is widely used and abundant throughout the world. Despite its significance to the global economy, its significance remains poorly understood. Presently, over 40 nations extract approximately 5 billion metric tons of minerals (Tropsch, 1929). Despite significant efforts and promotion of "clean fuels," coal remains the second most utilized energy source worldwide, following only oil. Based on estimates, it is projected that a quantity exceeding 5 billion tons is going to be necessary on a yearly basis by 2050, solely for the purpose of electricity generation. Coal basins have been regional sedimentary formations that exhibit a clustering for coal deposits within a particular area, commonly known as coal regions (Straka & Nahunkova, 2009). Globally, more than 2,000 sedimentary basins that bear coal are currently recognized, however, only a small fraction of these basins, specifically less than 12, possess reserves that exceed 200 billion tons. Nevertheless, the determination of coal reserves' quantity and lifespan (as with any natural resource's reserves) necessitates the inclusion of several assumptions (McCarthy et al., 2009).

- It is unlikely that any additional coal reserves are going to be discovered.
- Resources that are hypothetical or speculative in nature are unlikely to be classified as reserves.
- According to the stated conditions, there will likely be no emergence of novel mining technologies, thereby rendering conditional resources as non-reserves. Additionally, all coal that is accessible will be combusted, irrespective of its quality (Cameron et al., 1989).

Did You Know?

.The word coal comes from the old English term "col" which means charcoal

1.1. FUNDAMENTALS OF COAL

1.1.1. Coal Formation

Coal seems a type of sedimentary rock that is combustible that usually appears black in color. It is primarily composed of elements such as oxygen, hydrogen, along with carbon, with organic matter being the main source of its formation (Hutny et al., 1991).

Did You Know?

Coal ranks are often referred to as the coalification sequence. It represents the .progressive transformation of plant material into coal over millions of years

Fundamentally, coal comprises the transformed remnants of ancient flora that were initially situated in marshes and peatlands. Similar to all living entities, these botanical specimens accumulated energy via the sun via the mechanism of photosynthesis. Overall, the energy that is stored in plants is released through the process of decay as they perish. Occasionally, the natural process of decay was impeded by the accumulation of sediment along with other particulate matter, as well as geological shifts, which resulted in the entombment of plant matter at significant depths (Xiong et al., 2010). Over time, the subterranean organic matter experienced alterations in both its physical as well as chemical characteristics due to prolonged exposure to elevated temperatures and pressure, leading to its conversion into coal.

The coal development process commenced approximately 360 to 290 million years back, throughout the Carboniferous era, which is commonly referred to as the initial coal age. The intrinsic value of every coal deposit was determined by various factors such as temperature, pressure, and age. Initially, peat underwent a transformation process resulting in the formation of lignite, commonly known as 'brown coal' (Bhutto et al., 2013). Over the course of several millennia, the combination of heat and pressure has led to the transformation of lignite coal into a more energy-dense form known as sub-bituminous coal. Subsequent to initial alterations, additional chemical and physical processes took place, resulting in the development of increased density and darker pigmentation in the coal, ultimately giving rise to the classification of 'bituminous' or 'hard' coals (La Nauze, 1985). Under favorable circumstances, the gradual advancement in organic maturation persisted, ultimately resulting in the development of anthracite.

1.1.2. Coal Classification

The process by which coal transforms from lignite mining to anthracite can be described as "coalification." The classification of coal based on its chemical and physical attributes is commonly known as its "rank," which has significant implications for the characteristics of the coal (Alpern et al., 1989). The determination of ranking is based on the extent of conversion of the initial plant matter into carbon, as depicted in Figure 1.1.

Coals with a modest rank. Lignite as well as sub-bituminous coal exhibit a relatively low amount of carbon, yet possess a high proportion of hydrogen as well as oxygen, resulting in a reduced energy yield (Parr, 1922). Coal exhibiting a low rank is generally characterized by its softness, friability, and lackluster, earthy appearance.

Coal that has a high rank is characterized by a high carbon content, which results in a correspondingly elevated thermal value. Conversely, such coal exhibits fewer molecules of hydrogen as well as oxygen. In general, coal for higher rank exhibits greater hardness and strength and frequently displays a black, vitreous sheen (Norinaga et al., 1998). Anthracite ranks most prevalent upon the scale in terms of carbon as well as energy content while exhibiting the lowest amount of moisture.

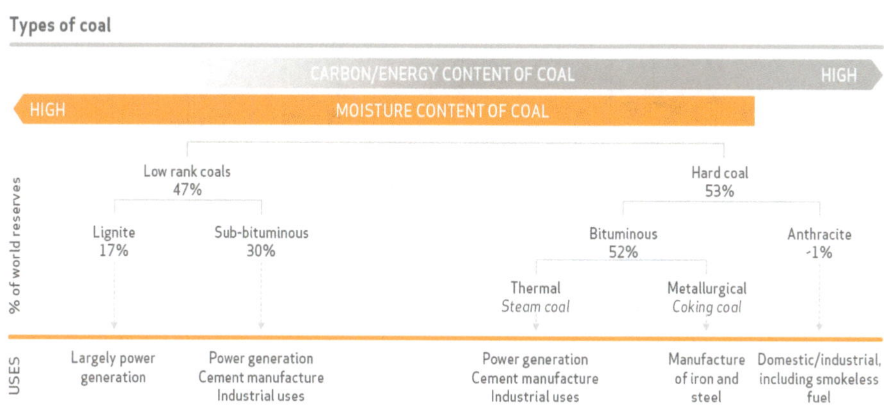

Figure 1.1. *The present study provides an illustration of various classifications of coal.*

Source: Liam McHugh, Creative Commons License.

Anthracite ranks most prevalent upon the scale in terms of carbon as well as energy content while exhibiting the lowest amount of moisture.

Did You Know?

Anthracite coal is sometimes referred to as "black diamond" due to its lustrous .appearance and high carbon content

1.1.3. Discovery of Coal

As per the report by the (International Energy Agency) IEA, the global proven coal reserves exceed 985 billion tons. Based on the present production rate, the available coal reserves have been projected to sustain for approximately 128 years (Friederich & van Leeuwen, 2017). Furthermore, it has been approximated that the existing coal reserves have the potential to meet the demand for an extended period, even beyond the decade that follows. This projection might be further prolonged through various advancements, such as:

- The identification of additional reserves via continuous and enhanced exploration endeavors.
- The progression of mining methodologies has enabled the exploitation of previously unattainable resources (Yang et al., 2017).

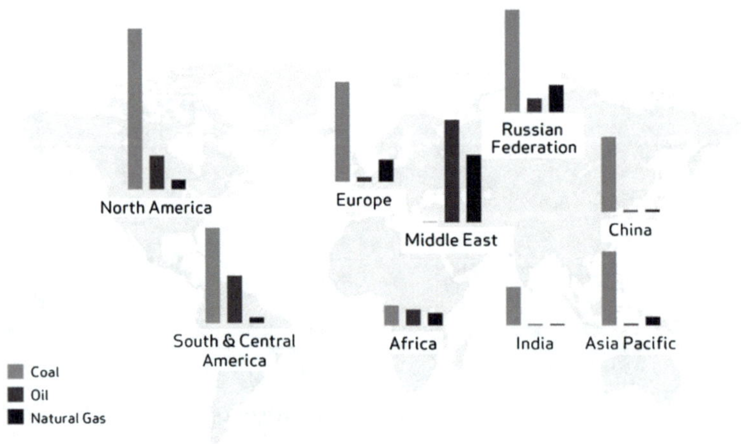

Figure 1.2. *The geographical distribution of the primary reserves for fossil fuels with the planet.*

Source: R. E. Hester, Creative Commons License.

As illustrated in Figure 1.2, reserves of coal are widely distributed across various regions, alongside recoverable reserves present in nearly 70 nations.

Coal is extracted in over 70 nations, despite the fact that the most extensive deposits are located in the United States, Russia, China, as well as India. Conversely, it is noteworthy that the Middle East region harbors nearly 50% of the global oil reserves, while Russia, Iran, along with Qatar collectively possess over 50% of the globe's gas reserves (Nyakuma et al., 2018).

The majority of coal is utilized for domestic purposes, with only a minor proportion of 17% being subject to international trading (Nyakuma, 2016). In several nations, coal represents the sole domestically accessible source of energy fuel, along with its utilization is incentivized by both economic as well as energy security rationales. This phenomenon is observed in various nations and territories, including Europe, China, along with India, where the reserves of coal surpass those of oil or gas. The majority of global coal exports are sourced from nations that are deemed politically stable, thereby mitigating the likelihood of supply disruptions.

1.1.4. Coal Exploration

The identification of coal reserves usually occurs through the implementation of exploration endeavors. Typically, the procedure entails generating a cartographic representation of the geological features present in the region, subsequently conducting assessments for the physical as well as chemical characteristics of the terrain, and ultimately executing exploratory drilling. This enables the development of a precise depiction of the region.

The feasibility of establishing a coal mine within a given area is contingent upon its size and quality, as only those meeting the requisite criteria for economic viability can be considered for extraction. Upon confirmation, the commencement of mining activities can ensue.

1.1.5. Coal Mining

There are two primary techniques employed in the extraction of coal, namely surface mining, also known to be opencast mining, and underground mining, which is also referred to as deep mining.

The selection of the mining technique is predominantly influenced by the geological characteristics of the coal reservoir (Zhengfu et al., 2010). At present, subterranean mining constitutes approximately 60% of global coal output, notwithstanding the fact that surface mining has become more prevalent in various significant coal-producing nations. Approximately 80% of the output within Australia is attributed to surface mining, whereas in America, this method accounts for approximately 67% of output.

1.1.5.1. Underground Mining

The subterranean extraction of minerals is typically accomplished through two primary techniques: room as well as pillar mining as well as longwall mining.

The extraction of coal deposits through the room, as well as the pillar mining method, entails the creation of a series of interconnected chambers within the coal seam while retaining columns of coal to provide structural reinforcement for the mine's roof (Hamrin et al., 2001). The coal pillars, which may constitute a significant proportion of the quantity of coal present in the seam, can occasionally be retrieved in a subsequent phase. The proportion of coal foundations can reach as much as 40%. The practice of extracting coal via pillars as miners retreat is commonly referred to as "retreat mining." Subsequently, the roof seems intentionally permitted to cave in, leading to the abandonment of the mine.

The process of longwall mineral extraction entails the complete retrieval of coal from a designated portion of the coal seam or just "face" through the utilization of mechanical shearers. Prior to commencing development work, it is imperative to meticulously plan a longwall encounter to ascertain the presence of favorable geology all over the section (Alford et al., 2007). The total length of the coal 'face' may range from 100 to 350 meters. During the coal extraction process, the roof is sustained by hydraulic-powered supports that have the ability to advance on their own. After the extraction of coal from the region, the roof is deliberately permitted to collapse. Approximately 75% or more of the coal present in the reserve can be obtained from coal panels that have the potential to extend up to 3 km via the coal seam.

The room along with pillar mining offers a significant advantage across longwall mining in terms of expedited coal production commencement, facilitated by the utilization of mobile equipment that incurs a relatively lower cost of beneath \$5 million, as opposed to longwall mining equipment that may cost up to \$50 million (Ramani, 2012).

The selection of mining methodology is contingent upon the unique characteristics of the site; however, it is invariably grounded in financial deliberations. Variations, even within a solitary mine, may necessitate the utilization of both techniques.

1.1.5.2. Surface Mining

Surface mining, additionally referred to by the terms opencast or just opencut mining, seems a viable option solely under the circumstance that the coal seam is situated in close proximity to the surface of the earth and can be extracted in a cost-effective manner. The approach utilized in this process enables the retrieval of a greater percentage of the coal reserve in comparison to subterranean mining, as it involves the exploitation of all coal seams, resulting in a recovery rate of at least 90% (Ercelebi & Bascetin, 2009). Opencast mining of considerable magnitude can span across numerous square kilometers and employ colossal machinery such as draglines for overburden removal, power shovels, mammoth trucks for overburden as well as coal transportation, bucket wheel digging machines, and conveyors.

The initial step in the sand and rock overload removal involves the utilization of explosives, followed by the extraction process through either draglines or shovel and truck methods (Mirmohammadi et al., 2009). Upon exposure of the coal seam, a drilling process is initiated, followed by controlled fracturing, as well as a systematic strip-mining approach is employed. Subsequently, the coal is transferred onto sizeable trucks or just conveyors for being conveyed to a combination of the coal prep facility or directly to its intended destination.

1.1.6. Coal Preparation

The coal within its natural state, referred to by the term run-of-mine coal, frequently harbors undesirable contaminants such as soil and stones and is comprised of a variety of fragment sizes. Nevertheless, consumers of coal necessitate a uniform level of coal (Noble & Luttrell, 2015). Coal preparing, which is also referred to as coal beneficiation or coal washing, pertains to the process of treating ROM coal in order to achieve uniform quality and improve its appropriateness for specific end applications.

The selection of an appropriate treatment method is contingent upon the specific characteristics associated with the coal as well as its intended application. The reduction of impurities in a substance may necessitate a straightforward crushing procedure or a more intricate treatment regimen. In order to eliminate contaminants, the unprocessed coal extracted from the mine is subjected to a crushing process and subsequently segregated into multiple size categories. The process of separating larger materials typically involves the utilization of "dense medium separation." The technique involves the separation of coal via other impurities through flotation in a tank that contains a liquid with significant specific gravity, typically a solution of finely crushed magnetite (Gui et al., 2015). Due to its lower density, coal exhibits buoyancy and can be segregated, whereas denser rock and extraneous substances settle and are eliminated as refuse. Various methods are employed for the treatment of smaller-size parts, typically relying on distinctions in mass, like those utilized in centrifugation. A centrifuge seems a mechanical device that rotates a vessel at high speeds, inducing the separation of solids as well as liquids within it. Diverse surface characteristics of coal as well as waste are exploited by alternative methods. The process of froth flotation involves the separation of coal particles through the generation of froth by introducing air through a bath of water that contains reagents made from chemicals. The phenomenon of selective adhesion of bubbles to transform coal fines, however rather than waste, is utilized in the process of recovering coal fines. Bubbles that are adhering to the coal fines have been subsequently separated and recovered (Peng et al., 2019). The retrieval of ultra-fine coal substances has been augmented by contemporary technological advancements.

1.1.7. Coal Transportation

The mode of transportation employed for coal delivery is contingent upon the extent of the path that needs that must be traversed. Typically, coal is conveyed or transported via truck for short distances. Trains as well as barges are utilized for extended distances within national markets (Kozan & Liu, 2012). On the other hand, coal may be combined with liquid water to create a coal slurry, which can be conveyed via a pipeline.

Ships are frequently utilized for global transportation as well as are available in a range of sizes, including Handymax (40–60,000 deadweight because tonnage or just DWT), Panamax (roughly 60–80,000 DWT), as well as big Capesize vessels (4,80,000 DWT).9 In the year 2015, the global international coal trading amounted to approximately 1,311 million tons (Mt), with the vast majority, approximately 90%, being conducted through seaborne transportation. The transporting of coal may result in significant expenses, with certain cases indicating that it constitutes as much as 70% of the total expense of coal delivery.

1.1.8. Coal Mining and the Environment

The environmental effects are mitigated throughout the entire process of coal shipping and storing through various measures.

The process of coal mining, especially surface mining, necessitates the temporary disturbance of vast tracts of land. The aforementioned circumstances pose several environmental obstacles, such as soil degradation, particulate matter dispersion, auditory disturbance, water contamination, and effects on indigenous flora and fauna (Goswami, 2015). Contemporary mining operations implement measures to mitigate the aforementioned impacts. Effective planning as well as environmental management strategies can mitigate the adverse effects of mining activities upon the environment, thereby contributing to the conservation of biodiversity.

1.1.8.1. Land Disturbance

As per established protocols, it is customary to conduct an assessment of the surrounding ecosystem several years prior to the commencement of coal mining operations. The purpose of this exercise is to ascertain the prevailing conditions as well as to identify any vulnerabilities or foreseeable issues that may arise (Xiao et al., 2020). The research investigates the effects of mining activities on surface and groundwater, soil quality, land utilization patterns, as well as the populations of indigenous flora and fauna. The utilization of computer simulations is a viable approach for modeling the potential impacts on the surrounding environment. The results are subsequently evaluated as a component of the procedure culminating in the issuance of a mining license by the pertinent governmental entities.

1.1.8.2. Mine Subsidence

Subsidence seems a prevalent issue that is often linked to subterranean coal mining, wherein the ground level experiences a reduction due to the extraction of coal from beneath the surface. Any use of land activity with the potential to endanger either private or public assets or valuable scenery is a matter of significant concern (Bahuguna et al., 1991). Comprehending subsistence patterns within a specific geographical area facilitates the quantification of the impact of subterranean mining upon the surface. This approach guarantees the secure and optimal retrieval of a coal reserve, while simultaneously safeguarding other land utilization.

1.1.8.3. Water Pollution

The phenomenon known as (acid mine drainage) AMD is characterized by the presence of water that is enriched with metals due to the reaction of chemicals among liquid and rocks that contain minerals carrying sulfur (Chaudhry & Malik, 2017). The runoff typically exhibits acidic properties and often originates from regions where mining operations involving the extraction of ore or coal have resulted in the exposure of pyrite-containing rocks, which are rich in sulfur. Nevertheless, the phenomenon of metal-rich drainage may also manifest in mineralized regions that haven't experienced any mining activities. The formation of AMD occurs as a result of the reaction between pyrite, air, as well as water, leading to the production of sulfuric acid as well as dissolved iron. The phenomenon of acid runoff entails the dissolution of heavy metals, including copper, lead, along with mercury, into both surface and groundwater. Various mine management techniques can mitigate the issue of AMD, while proficient mine design may successfully avert water from contacting acid-forming substances and thereby forestall the onset of AMD. The management of AMD may be approached through active or passive means (Freeman, 2010). The implementation of active treatment entails the installation of a water treatment facility. The acidic AMD is initially treated with lime to achieve neutralization, followed by the passage of the solution through obtaining tanks to eliminate sediment as well as particulate metals. The objective of passive treatment is to establish an autonomous system capable of treating the effluent lacking requiring continual human involvement.

1.1.8.4. Dust and Noise Pollution

In the context of mining activities, the implementation of contemporary mine planning methods as well as specialized equipment may serve to mitigate the effects of contaminants in the environment on both workers as well as local communities (Lilic et al., 2018). The generation of dust within mining activities can be attributed to several sources, including the operation of trucks on unpaved roads, the crushing of coal, drilling activities, and the dispersion of disturbed materials through the wind. The application of water on the roads, stockpiles, as well as conveyors, has been identified as a viable method for regulating dust levels. Further measures can be implemented to mitigate the impact of mining activities on the environment and surrounding communities. These may include the installation of dust collection systems on drilling equipment and the acquisition of adjacent land to serve as a protective barrier among the mine along with neighboring areas (Saha & Padhy, 2011). The planting of trees within these buffer zones has the potential to mitigate the visual effects of mining activities on nearby communities. Effective management of noise may be achieved by employing appropriate equipment, insulation, as well as sound enclosures within the vicinity of machinery. As per standard protocol, it is recommended to have a vibration as well as noise monitoring apparatus installed at every location to facilitate the measurement of noise levels as well as guarantee the mining activities are conducted within the prescribed limits.

1.1.9. Mine Rehabilitation

The temporary nature of coal mining necessitates the imperative of carrying out land rehabilitation post-cessation of mining activities (Wijesekara et al., 2016). It is considered to be ideal to formulate and obtain approval for a comprehensive rehabilitation or just reclamation plan associated with every coal mine, encompassing the duration from the commencement of operations till a considerable time after the conclusion of mining activities.

1.1.10. Mining Safety

The coal industry places a significant emphasis on the matter of safety. The extraction of coal from subterranean deposits poses a greater hazard to safety compared to coal extraction from open-air mines. Contemporary coal mines implement stringent safety protocols, uphold health and security regulations, and provide comprehensive training and instruction for workers (Karacan et al., 2011). These measures have resulted in notable enhancements in security measures for both subterranean and surface mining operations.

1.2. THE GLOBAL COAL MARKET

Coal finds its application in diverse sectors such as electricity generation, making steel and iron, and making cement, along with as a fluid fuel (Sribna et al., 2019). The predominant utilization of coal involves its application both in generating power, specifically steam coal or just lignite, or for manufacturing iron ore and steel, namely coking coal.

1.2.1. Coal Production

The current global coal production has exceeded 7,700 Mt, exhibiting a growth of 70% in the last two decades. The Asian region has experienced the most rapid growth in the mining of coal, whereas Europe has observed a decrease in manufacturing. The five-leading coal-producing nations, namely China, the United States, India, Australia, as well as Indonesia, are geographically diverse and not limited to a specific region.

It is projected that there will be a 4% rise in worldwide coal production until 2040, primarily driven by developing Asian countries which will account for nearly all of the augmented demand during this timeframe (Mohr & Evans, 2009). According to projections, the production of steam coal has reportedly attained a level of approximately 4.8 billion tons, while coking coal extraction surpassed 861 million tons.

1.2.2. Coal Consumption

Coal is a crucial component in the production of electricity and is expected to maintain its significance in the foreseeable future. At present, coal serves as the primary source of energy for 41% of global electricity production, and it is anticipated that this percentage will remain relatively stable in the upcoming three decades.

Asia represents the biggest consumer base for coal, comprising 75% of worldwide coal consumption. However, it is noteworthy that a considerable portion of this consumption can be attributed to China. Numerous nations lack adequate energy supplies to fulfill their energy demands, necessitating the importation of energy to supplement their requirements (Wolde-Rufael, 2010). Countries such as Japan, Chinese Taipei, as well as Korea, are known to import considerable amounts of steaming coal to generate electricity along with coking coal to produce steel.

The importation of coal by countries is driven not only by the scarcity of domestic indigenous coal reserves but also by the significance of acquiring particular coal varieties. Countries that are significant coal producers, including China, America, along with India, purchase coal for justifications related to quality and logistics. The utilization of coal is expected to persist as a significant component of the global energy portfolio, particularly in regions where there is a projected surge in demand (Li & Leung, 2012). The most robust expansion within the steam along with coking coal sectors is anticipated to occur in emerging Asian economies, where the escalation in electricity consumption and the requirement for steel within the construction, automobile manufacturing, along with household appliance sectors will intensify in tandem with income growth.

1.2.3. Coal Trade

Coal is a globally traded commodity; whereby vast distances are covered through maritime transportation to access various markets.

The maritime shipment of steam coal showed an average annual increase of approximately 17% throughout the past two decades, whereas the seaborne sale of coking coal showed a yearly increase of 3%. In 2015, the global coal trade amounted to 1,158 Mt. Despite the substantial volume, this figure represents a mere 15% of the total coal consumption (Haftendorn & Holz, 2010). A considerable proportion of the overall delivered value of coal relates to transportation expenses, leading to the division of the global steam coal business into two distinct regional markets, namely the Atlantic as well as the Pacific. The Atlantic market comprises nations within Western Europe that primarily engage in importing activities, with notable examples being the United Kingdom, Germany, as well as Spain. The Pacific marketplace encompasses both developing as well as OECD Asian buyers, with notable representation from Japan, Korea, along with Chinese Taipei. At present, the steam coal business is predominantly dominated by the Pacific market, which constitutes approximately 73% of all global trade (Ekawan et al., 2006). The phenomenon of market overlap is observed in situations where there is an abundance of coal supplies with elevated prices prevailing in the market. South Africa serves as the ideal convergence point for the two marketplaces.

1.3. UTILIZATION OF COAL

Coal plays a crucial role in facilitating modern society. Coal constitutes 41% of all global electricity supply and serves as a crucial primary material within the manufacturing of 70% of the globe's steel as well as 90% of the globe's cement (Bhattacharya & Kim, 2016).

1.3.1. Coal and Electricity

Electricity generated from coal can be generated by utilizing steam coal, which is also referred to as coal that is thermal. The initial step in the utilization of coal involves the milling of the substance into a finely powdered form. This process serves to augment the outermost area of the coal, thereby facilitating a more rapid combustion process (McNerney et al., 2011). Within pulverized coal-burning systems, finely ground coal is introduced into a boiler's burning chamber via a pneumatic transport system, where it is subjected to elevated temperatures and burned (*refer* Figure 1.3). The thermal energy and gaseous emissions generated during the combustion process facilitate the transformation of water, which is contained within the boiler's tubing, into steam.

The steam with high pressure flows into an engine that comprises numerous blades resembling propellers. The blades are propelled by steam, resulting in the rotation of the engine shaft at an elevated velocity (Lin & Jia, 2020). At one extremity of the engine shaft, a power source is installed. Upon traversing the turbine's blades, the steam undergoes condensation and subsequently re-enters the boiler for the purpose of reheating.

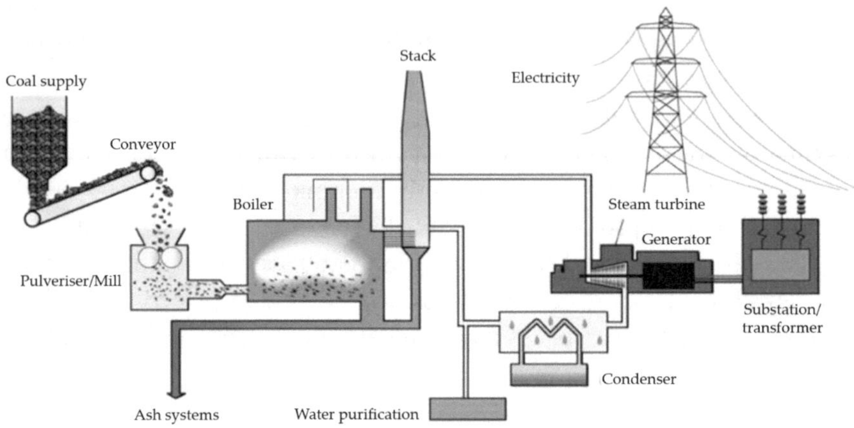

Figure 1.3. *The process of transforming coal into electrical energy.*

Source: R. M. Harrison, Creative Commons License.

The electrical energy produced is converted to high voltages, reaching as much as 400 kV, which are utilized for the purpose of cost-effective and efficient transmission through electrical grids. As power approaches the point of intake, namely residential households, it undergoes a transformation process to reduce its voltage to a more secure range of 100–250 V, which is commonly utilized within the domestic marketplace (Wang, 2007).

The availability of energy that is both cost-effective, dependable, and easily obtainable is a fundamental prerequisite for economic well-being in contemporary society.

Utilization of coal within Brazil has increased by more than 140% given that the 1980s, while India, as well as China, have experienced even more significant growth rates of 425% along with 514%, respectively. The recorded progress in both the financial and social domains across these nations during the aforementioned time frame is noteworthy. China has demonstrated a noteworthy illustration of the significance of cost-effective coal in enhancing energy accessibility and fostering economic advancement (Cui et al., 2012). As per the World Bank's estimations, approximately 600 million individuals have been alleviated from impoverished conditions over the past 30 years, with the vast majority of this progress occurring within China. Excluding China from the equation, poverty rates in other regions of the globe have exhibited minimal progress. The correlation between affordable coal-based power accessibility, economic expansion, as well as prosperity is evident. In China, nearly 99% of the populace has access to electricity through the national power grid.

1.3.2. Coal's Role in Delivering Modern Infrastructure

The utilization of energy-intensive content that includes steel, cement, transparent material, and aluminum is a crucial factor in determining the expansion of the economy (Liang et al., 2012). The utilization of these supplies is imperative for the establishment and progression of transportation, energy, residential, and water resource management infrastructure.

1.3.3. Steel Production

Coal plays a crucial role in the worldwide manufacturing of steel. The production of 1.3 tons of steel necessitates using a metric ton of coal for metallurgical purposes, equivalent to that.

The amount of energy required is comparable to the energy needed to produce 18 refrigerators at home. From 2001 to 2014, there has been a consistent upward trend in the per capita consumption of steel worldwide, with an average rise from 150 kg to 217 kg. This phenomenon reflects the increasing quality of life in developing nations and the resulting desire for consumer products (Tang et al., 2001).

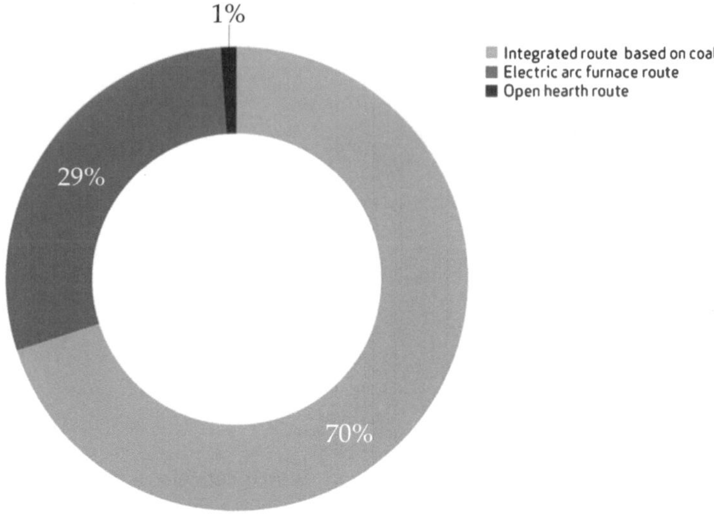

Figure 1.4. *Representation of crude steel production by process.*

Source: Liam McHugh, Creative Commons License.

The two primary steel-making methods are the incorporated steelmaking process and the electric arc furnace process. Coal is a fundamental resource in both contexts, as illustrated in Figure 1.4. The incorporated method, which relies on blast combustion and a basic oxygen combustion chamber, typically requires 770 kilograms of coal, 1,400 kilograms of iron ores, 150 kg of limestone, a rock, and 120 kg of reused steel per metric ton of crude steel produced (Muslemani et al., 2021). Typically, one metric ton of crude steel via the electrical arc furnace route necessitates the utilization of 880 kg of scrap steel, 150 kilograms of coal, and 43 kg of calcium carbonate.

Steel manufacturing is crucial in developing contemporary infrastructures, including transportation systems, residential dwellings, and commercial edifices. Steel production is instrumental in providing the necessary services and products essential for the functioning of a highly developed economy. These include but are not limited to healthcare, construction, and telecommuting, enhanced farming methods, improved transportation networks, accessibility to safe water, and dependable and reasonably priced energy.

1.3.4. Cement Production

Cement is a crucial component in manufacturing building materials, a fundamental construction material utilized in developing global social infrastructure second only (Schneider et al., 2011).

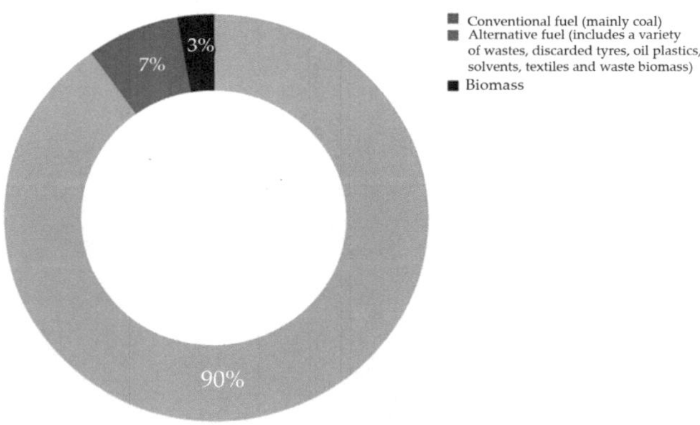

Figure 1.5. *Representation of total fuel consumption by cement.*

Source: R. E. Hester, Creative Commons License.

Next to water, it is the second most consumed substance in terms of quantity consumed annually. Cement is crucial in constructing various infrastructures such as residential buildings, bridges, highways, dams, harbors, or airfields. Coal is utilized as a primary energy source in making cement's basic components, such as limestone, rock silica, oxide of iron, and alumina, as depicted in Figure 1.5. The production of a metric ton of concrete requires approximately 200 kilograms of fossil fuels, and a cubic meter of concrete necessitates between 300 and 400 kilograms (Chen et al., 2010).

The Chinese market is a valuable indicator of the projected scale of cement consumption and, consequently, the requirement for coal in the foreseeable future. Between 2011 and 2013, China underwent a swift phase of industrialization and urbanization, during which it manufactured a quantity of cement that exceeded the total amount utilized by the United States throughout the 20th century (Habert et al., 2010).

1.3.5. Coal Liquefaction

Coal-derived petroleum products and coal-based power have the potential to make a substantial contribution to addressing the increasing energy demands of the transportation industry.

The production of liquid petroleum products to coal presents a feasible substitute for traditional oil-based products and has the potential to be integrated into the current supply facilities. China is currently in the process of developing multiple demonstration vegetation for coal-to-liquids (CTL) technology. The transport requirements of South Africa are partially met by CTL, which accounts for 20% of the total transportation needs (Shui et al., 2010). Additionally, CTL is responsible for supplying 7.5% of the jet fuel used in the country.

The process of coal, which involves the conversion of coal into liquid fuel, presents a viable option for substituting oil with coal. There exist two distinct techniques for transforming coal through fuel that is liquid which are discussed in subsections.

1.3.5.1. Direct Liquefaction

The process involves the dissolution of coal within a solvent under conditions of elevated pressure and elevated temperature (Huang & Yuan, 2015). The procedure above exhibits a notable degree of efficacy; however, additional refinement of the liquid byproducts is necessary to attain fuel properties of superior quality.

1.3.5.2. Indirect Liquefaction

The fossil fuel is subjected to gasification, forming syngas, a combination of carbon monoxide and hydrogen. Subsequently, the syngas undergoes condensation through a catalyst, commonly known as the 'Fischer-Tropsch' procedure, producing products of superior quality, and exhibiting exceptional cleanliness.

The liquid fuels produced from coal exhibit favorable characteristics such as the absence of sulfur, reduced particulate matter content, and minimal levels of nitrogen dioxide, thereby conferring advantages to the air quality of the area compared to petroleum-based fuels (Williams & Larson, 2003). Through the implementation of the co-treatment of coal along with biomass, coupled with carbon capture, utilization, and storage (CCUS), it is possible to achieve a reduction of up to 46% in emissions of carbon dioxide (CO_2) compared to liquid fossil fuels that originate from coal when compared to traditional oil-based goods, across the entirety of the power cycle.

1.3.6. Other Uses of Coal

Coal finds significant applications in various industries, such as alumina refineries, paper manufacturing, and biological and drug sectors. The waste products of coal have the potential to yield various chemical products. The production of various chemicals, including creosoting oil, Naphthalene, phenol, and benzene, involves using purified coal tar. The ammonia gas obtained from the coke ovens is utilized in producing ammonium salts, nitric acid, and fertilizers for agricultural purposes (Longwell et al., 1995). The coal and coal waste products are utilized as elements in many commodities, including but not limited to detergent, aspirin, chemicals, dyes, plastic bottles, and fibers like rayon and nylon. This results in incorporating coal into hundreds of thousands of unique consumer goods.

Coal is a crucial component in the manufacturing of specialized goods:

i. Activated carbon is a versatile material that finds application in various fields, such as air and water cleansing and dialysis for kidney machines.

ii. Carbon fiber is an additive widely used in various applications, such as bicycle construction and tennis balls, due to its exceptional strength and lightweight properties.

iii. Silicone metal is utilized in the production of silicones along with silanes, which are subsequently employed in the manufacturing of lubricants, in particular repellents that are resins that are beauty products, hair shampoos and conditioners, and dental products (Osborne & Gupta, 2013).

1.4. MEETING FUTURE ENERGY DEMAND

A significant obstacle currently confronting the global community is the need for access to contemporary energy services, which affects roughly 1.2 billion individuals, constituting 16% of the world's populace. According to Figure 1.6, an additional 2.7 billion individuals, which accounts for over 33% of the global population, rely on biomass, charcoal, or animal waste for warmth and cooking. In the past few years, there have been endeavors to stimulate action in this field (Al-Ahmad, 2018).

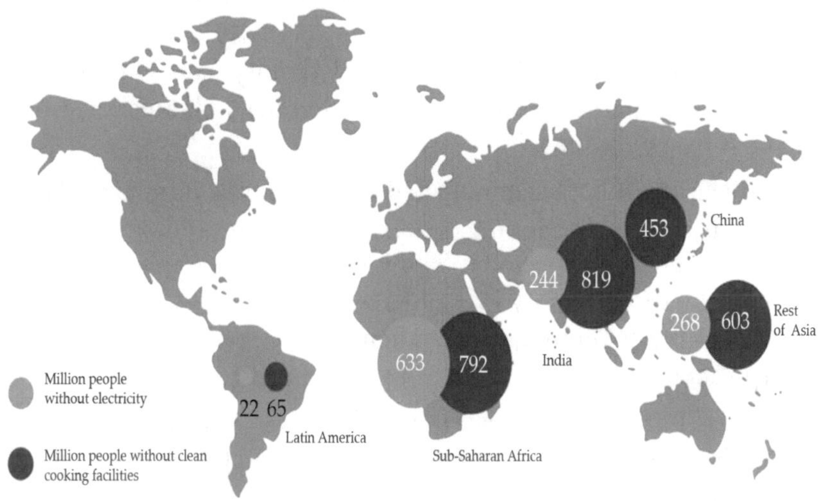

Figure 1.6. *Representation of people without access to modern energy services by region.*

Source: George R. M. Harrison, Creative Commons License.

Access to electricity has gained prominence in global development policy, as evidenced by various initiatives, including the Energy Access Action Strategy of the G20, the 2030 Agenda for Sustainable Development of the United Nations, and the Sustainable Energies for All program of the Secretary-General of the United Nations.

Although there have been some advancements in enhancing energy accessibility, the issue of maintaining a consistent and reliable energy supply stays a considerable obstacle. The attribution of this phenomenon by the IEA is based on a multitude of factors, which encompass the presence of intermittent energy sources in the capacity for producing electricity (Keles et al., 2017). These factors exhibit regional variability. Nevertheless, their implications are generally uniform.

Coal is a pragmatic option for furnishing consistent and widespread energy resources in numerous nations affected by energy insufficiency, owing to its extensive accessibility, dependability, and comparatively economical price. Due to this rationale, coal has emerged as the main energy source for electricity. According to research, from 1990 to 2010, a significant number of individuals, exceeding 750 million, predominantly residing in developing nations, could obtain electricity due to coal-fired power generation (Yuksek et al., 2006). The significance of coal in facilitating energy accessibility becomes more apparent when contrasted with periodic power sources. During the identical two-decade timeframe, scholarly investigations reveal that for each individual who obtained electricity accessibility through wind and solar means, approximately 13 individuals obtained accessibility through coal-based sources.

The utilization of coal is anticipated to increase by more than 11% until 2040, primarily due to developing nations' efforts to meet their electrification targets. Hence, coal is expected to significantly facilitate the establishment of base-load electricity supply in regions where it is highly essential, thereby contributing to the economic advancement of developing nations.

The IEA estimated that coal would be required to generate over 50% of the total on-grid electricity necessary to fulfill their 'energy for all' scenario, providing additional reinforcement to this prediction (Blaschke et al., 2013). It is noteworthy that the scenario above does not exhibit a particularly ambitious approach, as it only provides an output of just over five hours of power per day and does not encompass the provision of electricity for fundamental necessities such as businesses, industries, hospitals, schools, as well as government buildings.

1.4.1. Coal as an Important Element in the Balanced Energy Mix

According to the World Energy Microsoft Outlook (2016) released by the IEA, there is an anticipated increase of nearly one-third in worldwide demand for energy within the following 25 years. Most of the expansion above is anticipated in emerging economies, specifically in Asia's southern and southeast regions. Currently, the energy consumption per capita of an average Indian citizen is under a tenth of that of an average citizen belonging to the Organization for Economic Co-operation and Development (OECD) (Kaczmarek, 2022). It is inevitable that the progress of development within the region will result in a reduction of this disparity, which will demand a substantial augmentation in their dependence on power and travel and will necessitate the provision of significant additional energy resources.

To fulfill this requirement, the global community must acknowledge and utilize every possible energy source, such as coal. Despite the increasing prominence of renewable energy sources in the global energy portfolio, coal still holds a significant position in the energy landscape. In Germany, coal continues to serve as a source of base-load vitality for guaranteeing a reliable and safe power supply, despite a notable increase in the utilization of clean energy sources (Apostu et al., 2022).

Hence, energy participants must pursue the optimal approach to guarantee to attain both environmental and energy objectives. An optimal energy combination will incorporate diverse petroleum, nuclear power, and sporadic energy sources, utilizing centralized and decentralized representations. Indonesia serves as a definitive illustration in substantiating this proposition. The nation has pledged to augment its electricity capacity by 35 GW in the foreseeable future to support its economic expansion and tackle the electrification hurdles (Brodny & Tutak, 2022). The strategy entails the provision of low-cost electricity to densely populated and economically significant areas in Java through large, centralized coal-fired power plants. Meanwhile, off-grid renewable energy sources will cater to the energy needs of more remote regions in the archipelago. Per the World Economic Forum's recommendations, this methodology aims to reduce the costs associated with transmission and building, leverage native, and guarantee economical accessibility for all.

1.4.2. Coal as a Guarantor of Energy Security

The coal marketplace on a global scale is characterized by its vastness and heterogeneity, encompassing a multitude of makers and consumers from all corners of the world. The procurement of coal is not limited to a singular geographic location, thereby mitigating the risk of consumers being reliant on the security and stability of a solitary region (Mišík & Jursová Prachárová, 2023). Coal is traded globally and has a widespread distribution across the world. Based on the present mining rate, it is estimated that the existing coal reserves will be sufficient for approximately 110 years. When considering the current production rates, it is anticipated that the established reserves of natural gas and oil will endure for roughly 51 to 53 years, correspondingly.

Several nations, including China, the United States, India, Australia, and South Africa, depend on indigenous coal reserves to meet their power demands (Rybak & Rybak, 2018). Several countries engage in coal importation from diverse sources. For instance, in 2015, Germany purchased coal from a range of countries, including Russia, Colombia, and the United States, in addition to smaller quantities from various other nations and the company's domestic reserves.

Coal plays a crucial role in upholding the stability of the worldwide energy portfolio by serving as a supplement to other forms of energy and powers that are comparatively more susceptible to disturbances (Mouraviev, 2021). Coal plays a significant role in ensuring the stability and reliability of the electrical power mix through various means:

i. Reserves of coal are abundant and are expected to remain accessible for a while to come without giving rise to international or safety concerns.
ii. Coal is abundantly accessible from diverse sources in a globally well-stocked market.
iii. According to the statement, coal possesses the advantage of being conveniently storable at power plants and may be utilized as a reserve in unforeseen circumstances.
iv. Coal-fired power generation is characterized by its weather independence, making it a viable alternative to wind along with hydroelectricity to be additional (Sharples, 2012).
v. The fuel does not necessitate the utilization of high-pressure areas oil pipelines or exclusive transportation routes.
vi. The protection of coal transportation routes can be relatively inexpensive.

The characteristics mentioned above aid in promoting effective and competitive marketplaces for energy and contribute toward the price of energy.

1.4.3. On-Grid Electricity

The use of renewable energy has demonstrated notable efficacy in furnishing off-grid power supply to isolated communities in underdeveloped economies during recent years. The process of urbanization and industrialization necessitates establishing on-grid base-load power in nations. Renewable energy sources exhibit intermittency in their electricity generation, as the availability of wind and solar irradiance is subject to temporal fluctuations (Ma, 2011). The provision of base-load power is crucial for ensuring the dependability and consistency of electricity grids. Fossil fuels, such as coal, are significant sources of base-load power. Establishing base-load facilities facilitates the provision of advanced lifestyle amenities, such as refrigerators and cooling systems, while guaranteeing dependable electricity for commercial enterprises, medical facilities, public utilities, and industrial operations.

1.4.3.1. The Logical, Low-Cost, Base-Load Power Choice

The correlation between accessible coal-based power accessibility, job creation, and prosperity is evident (Forsberg, 2021). In the Indian context, it is observed that the cost of electricity generated from coal power plants is comparatively lower by 30% compared to electricity generated from renewable sources. It is also observed that the cost of electricity generated via coal-based strength plants is lower by 16% compared to domestic petroleum.

In the foreseeable future, coal will persist as the most cost-effective energy source for electricity production in numerous emerging nations for an extended period. The above-mentioned projection is substantiated by the examination outlined in The Energy Potential of High-Efficiency Coal; a report published by the World Coal Organization 2016. The study examines the levelized cost of electricity, or LCOE, across multiple technologies in non-OECD Asia. The LCOE offers a projected cost that power generators are expected to bear while generating electricity from diverse sources (Tian & You, 2021). The metric incorporates all anticipated expenses throughout the lifespan of a given system, encompassing construction, funding, fuel, upkeep, taxes, insurance, and rewards. These expenses are subsequently divided by the projected lifetime power output of the system, measured in kilowatt-hours (kWh). The computation process holds significant importance in formulating national energy production strategies as it facilitates the comparison of expenses associated with diverse technologies.

According to projections, in the year 2035, coal is anticipated to be a more cost-effective means of generating electricity in the Association of Southeast Asian Nations (ASEAN) countries, as well as India, Bangladesh, Sri Lanka, and Pakistan, in comparison to other methods, such as gas (Brown, 1981). The LCOE associated with high-efficiency, low-emissions (HELE) technologies, namely supercritical carbon dioxide (SC), ultra-supercritical coal (USC), and integrated combustion combined cycle (IGCC), exhibits a range of 55 to 60 dollars per megawatt-hour ($/MWh).

By contrast, the LCOE for open cycle gas turbines (OCGT) is nearly twice as high (Forsberg, 2021). It is noteworthy that in numerous nations, coal is more easily accessible than gas, necessitating the establishment of natural gas pipelines for transportation.

HELE technology exhibits a LCOE of approximately $50 per megawatt-hour, one-third of the cost of open-cycle gasoline turbines.

1.4.3.2. Low-Cost Electricity From Coal Aids Industrialization

The importance of stable and cost-effective energy pricing cannot be overstated for developing nations striving to establish their industrial capabilities, especially in sectors requiring significant energy consumption (Yi et al., 2015). The volatility of gasoline prices and elevated electricity expenses may result in a decline in a competitive edge and, in extended instances, the industry's demise.

As previously stated, the cost of electricity generated by power plants powered by coal in India is 30% lower than that generated from renewable sources (and 16% lower than domestic natural gas). The reduction in energy costs has discernible advantages for citizens. However, the impact of decreased energy prices is particularly evident for businesses. The reduction of energy costs significantly influences the promotion of industrial competitiveness (Shen et al., 2018). The reduced cost of electricity generated from coal leads to decreased production expenses, which enhances the profitability of the industrial sector and consequently contributes to the nation's economic growth. This, in turn, stimulates additional growth in the economy.

The recent establishment of the Sasan, the authority coal-fired nuclear plant in Madhya Pradesh, India, supports the theory. The power plant has contributed $2.5 billion to the regional economy by enhancing energy accessibility for nearby enterprises (Yang et al., 2015). The enhanced accessibility to energy has facilitated the deployment of electric pumps in the state's agricultural and related services sector. Enhanced accessibility to market information, including digital agricultural markets, has also proven advantageous for farmers. As mentioned, the advancements have improved efficiency, amplified yields, and increased crop variety.

An economy that needs more dependable energy sources may need help to achieve optimal performance. According to a survey by the International Monetary Fund, Indian business proprietors and senior executives were requested to choose the most substantial hindrance they face in their business, and power was found to be the second most major hurdle after corruption. In 2012, India experienced a blackout that lasted for two days and impacted 670 million individuals, accounting for over 10% of the global population (Bradu et al., 2022). The closure of shops, factories, and offices ensued, or they had to resort to backup generators, frequently powered by diesel. The repercussions on day-to-day commercial activities were experienced throughout the complete supply chain, as enterprises had to address not only disruptions in their premises but also those of their vendors. The Indian government took measures to enhance energy security by facilitating the establishment of extensive coal-based power generation facilities. The Sasan Authority Station plays a significant role in achieving this result. Madhya Pradesh, India's industrial sector, has experienced advantageous outcomes from a consistent energy supply, resulting in reduced losses attributed to power shortfalls by a sum exceeding $1.5 billion (Wrigley, 1962).

Enhancing productivity in both agricultural and industrial domains is paramount for a nation's economic expansion and development. According to a report published by the nation's National Council of Economic Research in 2013, creating a single job in the energy industry generates an additional 24.31 induced jobs in sectors such as commerce and agriculture (Wrigley, 2018). As illustrated in Figure 1.7, the projections indicate that coal will persist in producing electricity at a reasonable LCOE.

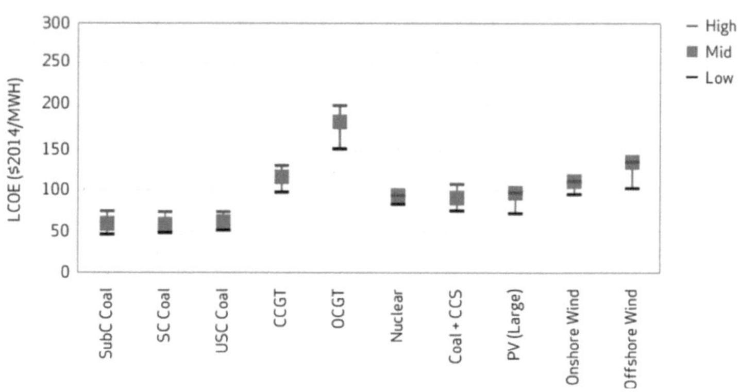

Figure 1.7. *Representation of India levelized cost of electricity by 2035.*

Source: Ben Caldecott, Creative Commons License.

1.4.3.3. Coal Sector Can Bring Broader Economic Benefits

The construction and operation of HELE coal power stations entail significant financial investments amounting to billions of dollars and necessitate the employment of a substantial workforce comprising thousands of individuals (Knowles & Ferbrache, 2016). The Coal Industry Advisory Board (CIAB) conducted an analysis revealing that more than 5,000 workers worked directly throughout the four-year building of a hypercritical facility in India. The facility's construction stimulated economic activity and resulted in the creation of an additional 3,700 indirect jobs. HELE plants, upon becoming operational, typically have life cycles spanning multiple decades and provide direct employment opportunities for professionals from diverse engineering disciplines, supervisors personnel involved in financial markets, management, employee relations, and security. The operationalization of HELE facilities leads to the continuation of indirect employment creation (Young et al., 2023). According to the CIAB analysis, approximately of employment opportunities, approximately 4,000, were generated through local agreements about the provision of labor, cleaning, gardening works, and vehicle availability.

Furthermore, as per the report's findings, the job opportunities generated by the establishment have resulted in a boost in economic activity throughout the neighboring community. Individuals directly or indirectly employed at the plant tended to possess greater financial freedom, leading to increased expenditures on consumer goods such as food, durable goods, and recreational activities (Steckel & Jakob, 2022). Based on the present analysis, it can be inferred that the facility in question has generated approximately 30,000 employment opportunities throughout its construction alongside various operational stages.

In addition, exploiting coal deposits for utilization in the domestic power industry confers economic advantages. Although coal extraction techniques may differ significantly among various areas, a shared attribute is that bitcoin mining is a costly operation that generates substantial economic revenue (LaBelle et al., 2023).

The coal-mining sector in South Africa, recognized as one of the leading manufacturers and consumers globally, constitutes over 17% of the country's mining labor force, employing more than 85,000 individuals. According to the Department of Mining Resources of the nation, the coal sector in South Africa remunerates its labor force with an amount exceeding US $1.2 billion annually (Spencer et al., 2018). The requisite expertise for roles in the coal mining industry may differ. However, a significant portion of the mechanical, technical, wellness, and safety instruction is imparted through on-the-job training.

Coal mining operations constitute a significant source of economic contribution for governmental entities. In 2014, the Republic of South Africa generated export revenue exceeding US $3 billion compared to coal, accounting for approximately 25% of the overall mining gross domestic product. The industry is a prominent source of government income through the collection of royalties. As per the National Treasury's Revenue estimations, these royalties amounted to a sum of US$ 400 million in the year 2013 (Lejour & De Mooij, 2005).

In recent times, the advancement of coal extraction has resulted in ancillary advantages via the creation of infrastructure. The infrastructure of rail lines along with roads, which are necessary for the transportation of fossil fuels to the mine/port, may be utilized by a diverse range of other industries. The Nacala port within Mozambique was initially established to export the nation's coal reserves and commenced its operations in 2015 (Gibson, 1984). Moreover, the port has facilitated the emergence of Mozambique as a pivotal transportation center in the region while also allowing coal export. Enhanced logistics have conferred advantages upon other adjacent economies, specifically Malawi, which is geographically landlocked and shares borders with Mozambique.

1.5. COAL AND THE ENVIRONMENT

Ecological requirements with the objectives of universal availability of energy, security of supply, and growth in society and the economy are crucial for the successful execution of the Agreement of Paris, which has now been formally adopted (Goswami, 2015). Achieving decreases in emissions in line with the goals of the Paris Accord and broader goals for development requires a delicate balance of the elements above.

Energy plays a crucial role in facilitating development. The availability of cost-effective and dependable electricity is fundamental to advancing success in contemporary society. As evidenced by the national contributions (NDCs), individual nations will select an optimal energy combination that aligns with their respective requirements (Bell & York, 2012). Numerous nations have recognized the persistent significance of coal, hence its continued utilization. This holds particularly true for the swiftly urbanizing and developing economies in Asia, in which coal is projected to serve as a crucial fuel source and is indispensable for economic expansion.

The initiative commences with implementing HELE energy production utilizing presently available technology (Li et al., 2022). The construction of these facilities is occurring swiftly, and they discharge a reduced amount of CO_2 by 25–33%. Additionally, they considerably diminish or eradicate the release of emissions, including nitrogen and sulfur oxides, as well as particulate matter.

Furthermore, implementing HELE technology signifies noteworthy advancement in the direction of CCUS, which is crucial for attaining worldwide climate targets (Li et al., 2022).

Acknowledging those above, numerous nations have incorporated a function for HELE coal-powered electricity production in their Intended Nationally Determined Contributions (INDCs). To facilitate the shift from outdated subcritical technology to more efficient alternatives, these nations will necessitate global financial, technological in nature, and other forms of assistance to expedite the implementation of such technology (Zhengfu et al., 2010).

1.5.1. Effciency Improvements

HELE technologies are currently available in the market as readily accessible products. These technologies are presently being implemented and utilized in numerous nations, yielding improvements in efficiency, and demonstrating financial feasibility (Abeykoon et al., 2021). Coal-fired power plants that utilize HELE technological advances have been found to emit 25–33% less CO_2 and considerably decrease or eliminate emissions of pollutants, including oxides of sulfur and nitrogen, as well as particulates. This is in contrast to older and less efficient subcritical technological advances.

The mean effectiveness of coal-fired power stations globally presently stands at 33%. The rate above falls significantly short of the current cutting-edge rate of 45% and even the readily available rates of approximately 40% (Menon et al., 2020). A 1% enhancement in the effectiveness of power plants powered by coal reduces CO_2 emissions ranging from 2 to 3%. The deployment of advanced technology to increase the efficiency rate of power stations powered by coal from the present worldwide median of 33% to 40% has the potential to reduce CO_2 emissions by 2 gigatons.

Contemporary patterns indicate that the correlation between the need for energy and the economy's advancement is progressively diminishing (Fong et al., 2020). The 2013 BP Energies Outlook predicts a reduction in energy magnitude, which refers to the quantity of energy needed per unit of GDP, by 36% (equivalent to an annual decrease of 1.9%) from 2012 to 2035. The progress above is a positive advancement that can be partially attributed to the participation of numerous nations in implementing HELE technologies, as opposed to outdated critical power plants powered by coal with lower efficiency.

However, there exists a potential danger of complacency after the ratification of the Paris Accord (Keevers & Green, 1993). In the forthcoming decades, developing nations must reconcile ecological imperatives, attain widespread energy access, ensure energy security, and promote social and economic progress. The absence of global monetary, technological nature and other assistance to expedite the implementation of HELE technology may impede the shift from sub-critical technology, undermining recent advancements in curbing energy intensity.

The 2015 WCA study, India's Energy Trilemma, analysis provides compelling evidence of the emission-reduction benefits of deploying HELE technologies (Wang et al., 2020). This analysis shows that replacing the subcritical capacity currently in India's development pipeline with supercritical or ultra-supercritical capacity could remove up to 11 billion tons of CO_2 emissions over the life of the power plants (*see* Figure 1.8). Adopting the most effcient HELE technology would abate the CO2 emissions equivalent to removing 500 million cars from the road.

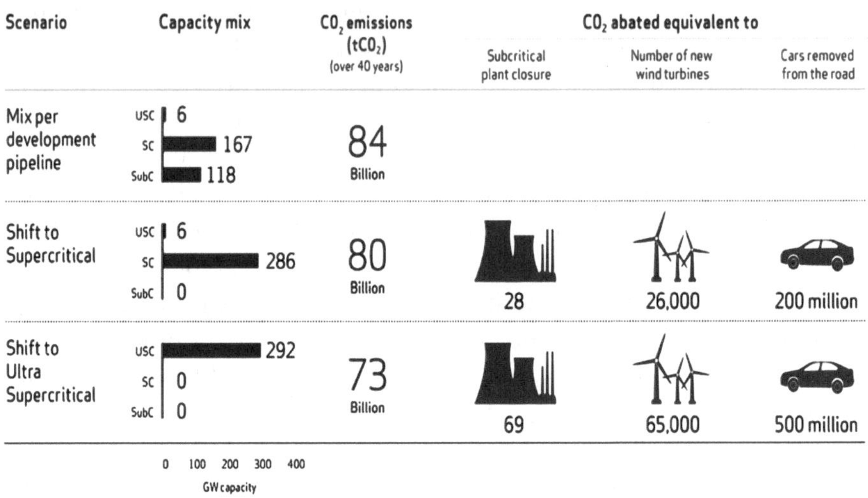

Figure 1.8. *Illustration of environmental benefits of deploying cleaner coal technology in India.*

Source: Zeshan Hyder, Creative Commons License.

The 2015 WCA research "India's Energy Trilemma" presents an analysis that offers convincing proof of the advantages of implementing

HELE methods in reducing emissions (Fourie, 2016). The present study demonstrates that substituting the critical capacity present in the developmental pipeline of India alongside hypercritical ultra-supercritical capacity can potentially eliminate nearly 11 billion tons of CO_2 throughout the lifespan of the power plants (*refer* Figure 1.8). Implementing the most optimal HELE technology has the potential to mitigate CO_2 greenhouse gases compared to removing 500 million automobiles from the roadways. The report highlights that despite the evident environmental ramifications, power plant builders may opt for less efficiency and inferior emissions rates owing to cost differentials, given the dearth of financing alternatives development banks offer (Phyu, 2009). A significant disparity of up to 40% exists in the capital expenditures of an ultra-supercritical fossil fuels plant and a critical thermal coal plant. Based on the analysis, it can be inferred that USC power plants exclusively for construction from 2020 onwards will result in total investment in the capital of $500 billion by 2040.

In contrast, if subcritical fossil fuel plants were exclusively constructed during the same period, the entire capital investment would come to approximately $387 billion. The 2016 report by WCA titled "The Power of High-Efficiency Coal" included supplementary analysis, which revealed that HELE technologies are often the most cost-effective option for reducing CO_2 emissions, based on a per ton basis (Prosser et al., 2011). The report analyzed the potential consequences of prioritizing assistance towards renewable energy sources while disregarding support for HELE coal-fired power generation. The research has revealed that the difficulty in augmenting the financial resources for implementing alternative energy generation lies in its low carry factor (Li et al., 2019). This implies that for every dollar invested, intermittent energy technologies replace significantly fewer MWh of subsequent subcritical generations than the HELE coal-fired generation. Consequently, in practical application, HELE coal-fired power generation reduces more CO_2 emissions than renewable energy sources per unit of investment. To provide contextualization, the IEA forecasts an estimated increase of 10,000 terawatt hours in electricity consumption in non-OECD Asian countries from 2020 to 2040 (Kiziroglou et al., 2016).

1.5.2. Carbon Capture, Utilization, and Storage (CCUS) Development Vital to Meeting Climate Goals

In light of society's continued dependence on fossil fuels, implementing CCUS technology is imperative to attain the necessary degree of emissions mitigation (Yan et al., 2021). Implementing CCUS technology is deemed necessary not just for coal but also for natural gas and industrial emissions to effectively mitigate the rise in global temperatures and maintain it below the 21C thresholds.

However, the present pace of CCUS implementation must be increased to attain the requisite emissions reduction targets. Hence, it is unreasonable to anticipate that the initial implementation of the deployment, as mentioned above, should be spearheaded by developing nations or burgeoning economies (Jiang et al., 2020). Rather than focusing on other aspects, the global community must prioritize the following three crucial domains to facilitate the implementation of CCUS.

1.5.2.1. International Financing for Carbon Capture, Utilization, and Storage (CCUS)

The accessibility of commercial financing is a significant obstacle to the expanded implementation of CCUS. CCUS is a capital-intensive technology that poses inherent risks, which tend to limit the availability of private financing. Although various initiatives have involved interactions with debt-finance suppliers, the Global CCS Institute (GCCSI) has indicated that the level of interest in and comprehension of CCS within the financial industry has not been extensively examined (Chen et al., 2022).

In such situations, global funding can offer crucial assistance. Several international funding organizations support the advancement of CCUS initiatives. These include the GCCSI, the Carbon Capture and Storage Training Fund of the World Bank, the Green Climate Fund, which was established by the CCUS Fund of the Asian Development Bank, and the Clean Development Mechanism of the United Nations Framework Convention on Climate Change (UNFCCC). However, extant programs can furnish only a portion of the necessary expenditure on capital (Jiang & Ashworth, 2021). The multilateral plans provide increased funding to enhance the financial feasibility of CCUS assignments.

The International Business Council for Sustainability, in collaboration with the Network for Sustainable Development Solutions and the IEA, has introduced the Low Carbon Dioxide Technology Partnerships Initiative (LCTPI), which presents a highly auspicious funding model (Kim et al., 2020). The platform above provides a venue for disseminating CCUS technology by eliminating obstacles and implementing necessary regulatory and economic mechanisms while advocating for collaborative efforts between governmental and private entities in technological advancement.

1.5.2.2. Policy Parity

Robust policies are instrumental in driving effective actions. The proliferation of green energy sources has been propelled by governmental policies that allocate $100 billion in annual subsidies (Morawski, 1994). The aggregate governmental policy assistance allocated to CCUS thus far constitutes roughly 1% of the total policy support extended to renewable energy technologies. The lack of availability of policy tools over CCUS compared to green power expenditure on the former. Providing robust policy backing for CCUS and equitable treatment alongside other low-emission technologies will facilitate the requisite financial commitment (Morawski, 1994). As mentioned above, the development will facilitate additional advancements in the broader exhibition and implementation of CCUS, consequently reducing expenses.

1.5.2.3. Carbon Capture, Utilization, and Storage (CCUS) Deployment Requires International Incentive

The implementation of climate change remedies necessitates concerted efforts on an international scale (Jiang & Ashworth, 2021). Sharing lessons learned from CCUS projects at global forums, such as the recently established USA-China Renewable Energy Research Center, is paramount. Furthermore, it is recommended that development financial institutions and donors establish financing mechanisms to support initiatives for research and development related to CCUS.

1.6. COAL AND OUR ENERGY FUTURE

The contemporary century poses numerous challenges to the global energy system (Clark et al., 2020). The challenge is to sustainably provide secure and affordable energy while simultaneously meeting the increasing demand. Simultaneously, there is a societal demand for less pollution and reduced emissions, coupled with a growing emphasis on preserving the environment (Bagge, 1977). Mitigating poverty, ensuring stable energy resources, and safeguarding the ecological system represent some of the most significant predicaments confronting our global community in the present era (Stiegel & Ramezan, 2006). Each of these difficulties is associated with the production and utilization of coal.

1.7. SUMMARY

Coal is a combustible black or brownish-black sedimentary rock that is primarily composed of carbon along with various other elements such as hydrogen, sulfur, oxygen, and nitrogen. In terms of its chemistry, coal is predominantly composed of carbon, which makes up a significant portion of its elemental composition. Coal is classified into different types or ranks based on its composition, properties, and carbon content. The classification of coal is primarily determined by the degree of coalification, which is the extent to which the original plant material has been transformed into coal through geological processes. The study of coal chemistry is important for understanding its properties, as well as for developing cleaner and more efficient methods for its use.

REFERENCES

1. Abeykoon, C., McMillan, A., & Nguyen, B. K., (2021). Energy efficiency in extrusion-related polymer processing: A review of state of the art and potential efficiency improvements. *Renewable and Sustainable Energy Reviews*, *147*, 111219.
2. Al-Ahmad, H., (2018). Biotechnology for bioenergy dedicated trees: Meeting future energy demands. *Zeitschrift für Naturforschung C*, *73*(1–2), 15–32.
3. Alford, C., Brazil, M., & Lee, D. H., (2007). Optimization in underground mining. *Handbook of Operations Research in Natural Resources*, *1*, 561–577.
4. Alpern, B., De Sousa, M. L., & Flores, D., (1989). A progress report on the Alpern coal classification. *International Journal of Coal Geology*, *13*(1–4), 1–19.
5. Apostu, I. M., Lazar, M., & Faur, F., (2022). Implications of coal disappearance from the energy mix in Romania. In: *MATEC Web of Conferences* (Vol. 354, p. 00057).
6. Bagge, C. E., (1977). Coal and the nation's energy future. *Hous. L. REv.*, *15*, 1081.
7. Bahuguna, P. P., Srivastava, A. M. C., & Saxena, N. C., (1991). A critical review of mine subsidence prediction methods. *Mining Science and Technology*, *13*(3), 369–382.
8. Bell, S. E., & York, R., (2012). Coal, injustice, and environmental destruction: Introduction to the special issue on coal and the environment. *Organization and Environment*, *25*(4), 359–367.
9. Bhattacharya, S. S., & Kim, K. H., (2016). Utilization of coal ash: Is vermitechnology a sustainable avenue? *Renewable and Sustainable Energy Reviews*, *58*, 1376–1386.
10. Bhutto, A. W., Bazmi, A. A., & Zahedi, G., (2013). Underground coal gasification: From fundamentals to applications. *Progress in Energy and Combustion Science*, *39*(1), 189–214.
11. Blaschke, T., Biberacher, M., Gadocha, S., & Schardinger, I., (2013). 'Energy landscapes': Meeting energy demands and human aspirations. *Biomass and Bioenergy*, *55*, 3–16.
12. Bradu, P., Biswas, A., Nair, C., Sreevalsakumar, S., Patil, M., Kannampuzha, S., & Gopalakrishnan, A. V., (2022). Recent advances in green technology and industrial revolution 4.0 for a sustainable future. *Environmental Science and Pollution Research*, 1–32.

13. Brodny, J., & Tutak, M., (2022). Challenges of the polish coal mining industry on its way to innovative and sustainable development. *Journal of Cleaner Production*, *375*, 134061.
14. Brown, W. C., (1981). The solar power satellite as a source of base load electrical power. *IEEE Transactions on Power Apparatus and Systems*, *6*, 2766–2774.
15. Cameron, C. C., Esterle, J. S., & Palmer, C. A., (1989). The geology, botany, and chemistry of selected peat-forming environments from temperate and tropical latitudes. *International Journal of Coal Geology*, *12*(1–4), 105–156.
16. Chaudhry, F. N., & Malik, M. F., (2017). Factors affecting water pollution: a review. *J. Ecosyst. Ecography*, *7*(1), 225–231.
17. Chen, C., Habert, G., Bouzidi, Y., & Jullien, A., (2010). Environmental impact of cement production: detail of the different processes and cement plant variability evaluation. *Journal of Cleaner Production*, *18*(5), 478–485.
18. Chen, S., Liu, J., Zhang, Q., Teng, F., & McLellan, B. C., (2022). A critical review on deployment planning and risk analysis of carbon capture, utilization, and storage (CCUS) toward carbon neutrality. *Renewable and Sustainable Energy Reviews*, *167*, 112537.
19. Chiesa, P., Consonni, S., Kreutz, T., & Williams, R., (2005). Co-production of hydrogen, electricity, and CO_2 from coal with commercially ready technology. Part A: Performance and emissions. *International Journal of Hydrogen Energy*, *30*(7), 747–767.
20. Clark, R., Zucker, N., & Urpelainen, J., (2020). The future of coal-fired power generation in Southeast Asia. *Renewable and Sustainable Energy Reviews*, *121*, 109650.
21. Cui, X., Hong, J., & Gao, M., (2012). Environmental impact assessment of three coal-based electricity generation scenarios in China. *Energy*, *45*(1), 952–959.
22. Davis, K. J., & Gerlach, R., (2018). Transition of biogenic coal-to-methane conversion from the laboratory to the field: A review of important parameters and studies. *International Journal of Coal Geology*, *185*, 33–43.
23. Ekawan, R., Duchêne, M., & Goetz, D., (2006). The evolution of hard coal trade in the Pacific market. *Energy Policy*, *34*(14), 1853–1866.
24. Ercelebi, S. G., & Bascetin, A., (2009). Optimization of shovel-truck system for surface mining. *Journal of the Southern African Institute of Mining and Metallurgy*, *109*(7), 433–439.

25. Fong, A. J., Smith, M., & Langerman, A., (2016). Efficiency improvement in the operating room. *Journal of Surgical Research*, *204*(2), 371–383.
26. Forsberg, C., (2021). 100-gigawatt-hour crushed-rock heat storage for variable electricity and heat with base-load reactor operations. In: *International Conference on Nuclear Engineering* (Vol. 85246, p. V001T02A027).
27. Fourie, H., (2016). Improvement in the overall efficiency of mining equipment: A case study. *Journal of the Southern African Institute of Mining and Metallurgy*, *116*(3), 275–281.
28. Freeman III, A. M., (2010). Water pollution policy. In: *Public Policies for Environmental Protection,* (Vol. 1, pp. 179–224).
29. Friederich, M. C., & van Leeuwen, T., (2017). A review of the history of coal exploration, discovery, and production in Indonesia: The interplay of legal framework, coal geology and exploration strategy. *International Journal of Coal Geology*, *178*, 56–73.
30. Friedmann, S. J., Upadhye, R., & Kong, F. M., (2009). Prospects for underground coal gasification in carbon-constrained world. *Energy Procedia*, *1*(1), 4551–4557.
31. Gibson, K. D., (1984). Industrial reorganization and coal production in Australia, 1860–1982: An historical materialist analysis. *Australian Geographical Studies*, *22*(2), 221–242.
32. Goswami, S., (2015). Impact of coal mining on environment. *European Researcher*, *3*, 185–196.
33. Gui, X., Liu, J., Cao, Y., Miao, Z., Li, S., Xing, Y., & Wang, D., (2015). Coal preparation technology: Status and development in China. *Energy and Environment*, *26*(6–7), 997–1013.
34. Habert, G., Billard, C., Rossi, P., Chen, C., & Roussel, N., (2010). Cement production technology improvement compared to factor 4 objectives. *Cement and Concrete Research*, *40*(5), 820–826.
35. Haftendorn, C., & Holz, F., (2010). Modeling and analysis of the international steam coal trade. *The Energy Journal*, *31*(4).
36. Hamrin, H., Hustrulid, W., & Bullock, R., (2001). Underground mining methods and applications. *Underground Mining Methods: Engineering Fundamentals and International Case Studies*, *1*(3–14).
37. Huang, H. J., & Yuan, X. Z., (2015). Recent progress in the direct liquefaction of typical biomass. *Progress in Energy and Combustion Science*, *49*, 59–80.
38. Hutny, W. P., Lee, G. K., & Price, J. T., (1991). Fundamentals of coal

combustion during injection into a blast furnace. *Progress in Energy and Combustion Science*, *17*(4), 373–395.

39. Jiang, K., & Ashworth, P., (2021). The development of carbon capture utilization and storage (CCUS) research in China: A bibliometric perspective. *Renewable and Sustainable Energy Reviews*, *138*, 110521.
40. Jiang, K., & Ashworth, P., (2021). The development of carbon capture utilization and storage (CCUS) research in China: A bibliometric perspective. *Renewable and Sustainable Energy Reviews*, *138*, 110521.
41. Jiang, K., Ashworth, P., Zhang, S., Liang, X., Sun, Y., & Angus, D., (2020). China's carbon capture, utilization, and storage (CCUS) policy: A critical review. *Renewable and Sustainable Energy Reviews*, *119*, 109601.
42. Kaczmarek, J., (2022). The balance of outlays and effects of restructuring hard coal mining companies in terms of energy policy of Poland PEP 2040. *Energies*, *15*(5), 1853.
43. Karacan, C. Ö., Ruiz, F. A., Cotè, M., & Phipps, S., (2011). Coal mine methane: A review of capture and utilization practices with benefits to mining safety and to greenhouse gas reduction. *International Journal of Coal Geology*, *86*(2–3), 121–156.
44. Keevers, M. J., & Green, M. A., (1993). Efficiency improvements of silicon solar cells by the impurity photovoltaic effect. In: *Conference Record of the 23rd IEEE Photovoltaic Specialists Conference–1993 (Cat. No. 93CH3283–9)* (pp. 140–146).
45. Keles, D., Jochem, P., McKenna, R., Ruppert, M., & Fichtner, W., (2017). Meeting the modeling needs of future energy systems. *Energy Technology*, *5*(7), 1007–1025.
46. Kim, I., Yoo, Y., Son, J., Park, J., Huh, I. S., & Kang, D., (2020). Two-step mineral carbonation using seawater-based industrial wastewater: An eco-friendly carbon capture, utilization, and storage process. *Journal of Material Cycles and Waste Management*, *22*, 333–347.
47. Kiziroglou, M. E., Boyle, D. E., Yeatman, E. M., & Cilliers, J. J., (2016). Opportunities for sensing systems in mining. *IEEE Transactions on Industrial Informatics*, *13*(1), 278–286.
48. Knowles, R. D., & Ferbrache, F., (2016). Evaluation of wider economic impacts of light rail investment on cities. *Journal of Transport Geography*, *54*, 430–439.
49. Kozan, E., & Liu, S. Q., (2012). A demand-responsive decision support system for coal transportation. *Decision Support Systems*, *54*(1), 665–680.
50. Kreutz, T., Williams, R., Consonni, S., & Chiesa, P., (2005). Co-production

of hydrogen, electricity, and CO_2 from coal with commercially ready technology. Part B: Economic analysis. *International Journal of Hydrogen Energy*, *30*(7), 769–784.

51. La Nauze, R. D., (1985). Fundamentals of coal combustion in fluidized beds. *Chem. Eng. Res. Des. (United Kingdom)*, *63*(1).
52. LaBelle, M. C., Bucată, R., & Stojilovska, A., (2023). Radical energy justice: A Green Deal for Romanian coal miners? *Journal of Environmental Policy and Planning*, *25*(2), 142–154.
53. Lejour, A. M., & De Mooij, R. A., (2005). Turkish delight: Does Turkey's accession to the EU bring economic benefits? *Kyklos*, *58*(1), 87–120.
54. Li, R., & Leung, G. C., (2012). Coal consumption and economic growth in China. *Energy Policy*, *40*, 438–443.
55. Li, Y., Chiu, Y. H., & Lin, T. Y., (2019). Coal production efficiency and land destruction in China's coal mining industry. *Resources Policy*, *63*, 101449.
56. Li, Y., Pan, S., Ning, S., Shao, L., Jing, Z., & Wang, Z., (2022). Coal measure metallogeny: Metallogenic system and implication for resource and environment. *Science China Earth Sciences*, *65*(7), 1211–1228.
57. Liang, F. Y., Ryvak, M., Sayeed, S., & Zhao, N., (2012). The role of natural gas as a primary fuel in the near future, including comparisons of acquisition, transmission, and waste handling costs of as with competitive alternatives. *Chemistry Central Journal*, *6*(1), 1–24.
58. Lilic, N., Cvjetic, A., Knezevic, D., Milisavljevic, V., & Pantelic, U., (2018). Dust and noise environmental impact assessment and control in Serbian mining practice. *Minerals*, *8*(2), 34.
59. Lin, B., & Jia, Z., (2020). Economic, energy, and environmental impact of coal-to-electricity policy in China: A dynamic recursive CGE study. *Science of the Total Environment*, *698*, 134241.
60. Liu, H., & Liu, S., (2021). Life cycle energy consumption and GHG emissions of hydrogen production from underground coal gasification in comparison with surface coal gasification. *International Journal of Hydrogen Energy*, *46*(14), 9630–9643.
61. Longwell, J. P., Rubin, E. S., & Wilson, J., (1995). Coal: Energy for the future. *Progress in Energy and Combustion Science*, *21*(4), 269–360.
62. Ma, J., (2011). On-grid electricity tariffs in China: Development, reform, and prospects. *Energy Policy*, *39*(5), 2633–2645.
63. McCarthy, T. S., McIver, J. R., Cairncross, B., Ellery, W. N., & Ellery, K., (1990). The inorganic chemistry of peat from the Okavango delta (fan),

Botswana: implications for low-ash coal formation. *International Journal of Coal Geology*, *16*(1–3), 143–146.

64. McNerney, J., Farmer, J. D., & Trancik, J. E., (2011). Historical costs of coal-fired electricity and implications for the future. *Energy Policy*, *39*(6), 3042–3054.
65. Menon, A., Stojceska, V., & Tassou, S. A., (2020). A systematic review on the recent advances of the energy efficiency improvements in non-conventional food drying technologies. *Trends in Food Science and Technology*, *100*, 67–76.
66. Midilli, A., Kucuk, H., Topal, M. E., Akbulut, U., & Dincer, I., (2021). A comprehensive review on hydrogen production from coal gasification: Challenges and Opportunities. *International Journal of Hydrogen Energy*, *46*(50), 25385–25412.
67. Mirmohammadi, M., Gholamnejad, J., Fattahpour, V., Seyedsadri, P., & Ghorbani, Y., (2009). Designing of an environmental assessment algorithm for surface mining projects. *Journal of Environmental Management*, *90*(8), 2422–2435.
68. Mišík, M., & Jursová Prachárová, V., (2023). Coal is a priority for energy security, until it is not: coal phase-out in the EU and its persistence in the face of the energy crisis. *Sustainability*, *15*(8), 6879.
69. Mohr, S. H., & Evans, G. M., (2009). Forecasting coal production until 2100. *Fuel*, *88*(11), 2059–2067.
70. Morawski, E., (1994). Restructuring the hard coal mining industry in Poland. *Resources Policy*, *20*(3), 193–197.
71. Mouraviev, N., (2021). Energy security in Kazakhstan: The consumers' perspective. *Energy Policy*, *155*, 112343.
72. Muslemani, H., Liang, X., Kaesehage, K., Ascui, F., & Wilson, J., (2021). Opportunities and challenges for decarbonizing steel production by creating markets for 'green steel' products. *Journal of Cleaner Production*, *315*, 128127.
73. Noble, A., & Luttrell, G. H., (2015). A review of state-of-the-art processing operations in coal preparation. *International Journal of Mining Science and Technology*, *25*(4), 511–521.
74. Norinaga, K., Kumagai, H., Hayashi, J. I., & Chiba, T., (1998). Classification of water sorbed in coal on the basis of congelation characteristics. *Energy and Fuels*, *12*(3), 574–579.
75. Nyakuma, B. B., Jauro, A., Oladokun, O., Bello, A. A., Alkali, H., Modibo, M. A., & Abba, M., (2018). Physicochemical, mineralogical,

and thermogravimetric properties of newly discovered Nigerian coals. *Petroleum and Coal*, *60*(4).

76. Nyakuma, B., (2016). Physicochemical characterization and thermal analysis of newly discovered Nigerian coals. *Bulgarian Chemical Communications*, *48*(4), 746–752.
77. Orem, W. H., & Finkelman, R. B., (2003). *Coal Formation and Geochemistry* (Vol. 7, p. 407).
78. Osborne, D. G., & Gupta, S. K., (2013). Industrial uses of coal. In: *The Coal Handbook: Towards Cleaner Production* (Vol. 1, pp. 3–30).
79. Parr, S. W., (1922). The classification of coal. *Industrial and Engineering Chemistry*, *14*(10), 919–922.
80. Peng, L., Jiang, H., Chen, X., Liu, D., Feng, H., Zhang, L., & Liu, C., (2019). A review on the advanced design techniques and methods of vibrating screen for coal preparation. *Powder Technology*, *347*, 136–147.
81. Phyu, T. N., (2009). Survey of classification techniques in data mining. In: *Proceedings of the International Multiconference of Engineers and Computer Scientists* (Vol. 1, No. 5, pp. 727–731).
82. Prosser, I., Wolf, L., & Littleboy, A., (2011). Water in mining and industry. *Water: Science and Solutions for Australia*, 135–146.
83. Ramani, R. V., (2012). Surface mining technology: progress and prospects. *Procedia Engineering*, *46*, 9–21.
84. Rybak, A., & Rybak, A., (2018). The future of hard coal as a guarantor of energy security in Poland. *International Multidisciplinary Scientific GeoConference: SGEM*, *18*(4.1), 681–688.
85. Saha, D. C., & Padhy, P. K., (2011). Effect of air and noise pollution on species diversity and population density of forest birds at Lalpahari, West Bengal, India. *Science of the Total Environment*, *409*(24), 5328–5336.
86. Schneider, M., Romer, M., Tschudin, M., & Bolio, H., (2011). Sustainable cement production – present and future. *Cement and Concrete Research*, *41*(7), 642–650.
87. Schopf, J. M., (1948). Variable coalification; the processes involved in coal formation. *Economic Geology*, *43*(3), 207–225.
88. Scott, A. C., (2002). Coal petrology and the origin of coal macerals: A way ahead? *International Journal of Coal Geology*, *50*(1–4), 119–134.
89. Sharples, J., (2012). Russo–Polish energy security relations: A case of threatening dependency, supply guarantee, or regional energy security dynamics? *Political Perspectives*, *6*(1), 27–50.
90. Shen, X., Tian, Z., Fan, R., Shao, L., Zhang, D., Cao, G., & Bai, Y.,

(2018). Research progress on silicon/carbon composite anode materials for lithium-ion battery. *Journal of Energy Chemistry*, *27*(4), 1067–1090.

91. Shoko, E., McLellan, B., Dicks, A. L., & Da Costa, J. D., (2006). Hydrogen from coal: Production and utilization technologies. *International Journal of Coal Geology*, *65*(3–4), 213–222.
92. Shui, H., Cai, Z., & Xu, C., (2010). Recent advances in direct coal liquefaction. *Energies*, *3*(2), 155–170.
93. Spencer, T., Colombier, M., Sartor, O., Garg, A., Tiwari, V., Burton, J., & Wiseman, J., (2018). The 1.5 C target and coal sector transition: At the limits of societal feasibility. *Climate Policy*, *18*(3), 335–351.
94. Sribna, Y., Trokhymets, O., Nosatov, I., & Kriukova, I., (2019). The globalization of the world coal market-contradictions and trends. In: *E3S Web of Conferences* (Vol. 123, p. 01044).
95. Steckel, J. C., & Jakob, M., (2022). To end coal, adapt to regional realities. *Nature*, *607*(7917), 29–31.
96. Stiegel, G. J., & Ramezan, M., (2006). Hydrogen from coal gasification: An economical pathway to a sustainable energy future. *International Journal of Coal Geology*, *65*(3–4), 173–190.
97. Straka, P., & Nahunkova, J., (2009). Temperature conditions of coal formation. *Acta Geodyn. Geomater*, *6*(154), 201–207.
98. Tang, L., Liu, J., Rong, A., & Yang, Z., (2001). A review of planning and scheduling systems and methods for integrated steel production. *European Journal of Operational Research*, *133*(1), 1–20.
99. Tian, X., & You, F., (2021). Sustainable design of hybrid energy systems towards carbon neutrality. *Chemical Engineering Transactions*, *88*, 1207–1212.
100. Tropsch, H., (1929). Problems in the Chemistry of Coal. *Chemical Reviews*, *6*(1), 63–90.
101. Wang, B., (2007). An imbalanced development of coal and electricity industries in China. *Energy Policy*, *35*(10), 4959–4968.
102. Wang, Y., Lei, Y., & Wang, S., (2020). Green mining efficiency and improvement countermeasures for China's coal mining industry. *Frontiers in Energy Research*, *8*, 18.
103. Wijesekara, H., Bolan, N. S., Vithanage, M., Xu, Y., Mandal, S., Brown, S. L., & Surapaneni, A., (2016). Utilization of biowaste for mine spoil rehabilitation. *Advances in Agronomy*, *138*, 97–173.
104. Williams, R. H., & Larson, E. D., (2003). A comparison of direct and indirect liquefaction technologies for making fluid fuels from coal. *Energy*

for Sustainable Development, *7*(4), 103–129.

105. Wolde-Rufael, Y., (2010). Coal consumption and economic growth revisited. *Applied Energy*, *87*(1), 160–167.
106. Wrigley, E. A., (1962). The supply of raw materials in the industrial revolution. *The Economic History Review*, *15*(1), 1–16.
107. Wrigley, E. A., (2018). Reconsidering the industrial revolution: England and Wales. *Journal of Interdisciplinary History*, *49*(1), 9–42.
108. Xiao, W., Deng, X., He, T., & Chen, W., (2020). Mapping annual land disturbance and reclamation in a surface coal mining region using google earth engine and the Landtrendr algorithm: A case study of the Shengli coalfield in Inner Mongolia, China. *Remote Sensing*, *12*(10), 1612.
109. Xiong, R., Dong, L., Yu, J., Zhang, X., Jin, L., & Xu, G., (2010). Fundamentals of coal topping gasification: Characterization of pyrolysis topping in a fluidized bed reactor. *Fuel Processing Technology*, *91*(8), 810–817.
110. Yan, Y., Borhani, T. N., Subraveti, S. G., Pai, K. N., Prasad, V., Rajendran, A., & Clough, P. T., (2021). Harnessing the power of machine learning for carbon capture, utilization, and storage (CCUS)–a state-of-the-art review. *Energy and Environmental Science*, *14*(12), 6122–6157.
111. Yang, Y., Chen, B., Hower, J., Schindler, M., Winkler, C., Brandt, J., & Hochella Jr, M. F., (2017). Discovery and ramifications of incidental Magnéli phase generation and release from industrial coal-burning. *Nature Communications*, *8*(1), 194.
112. Yang, Z., Ren, J., Zhang, Z., Chen, X., Guan, G., Qiu, L., & Peng, H., (2015). Recent advancement of nanostructured carbon for energy applications. *Chemical Reviews*, *115*(11), 5159–5223.
113. Yi, Q., Li, W., Feng, J., & Xie, K., (2015). Carbon cycle in advanced coal chemical engineering. *Chemical Society Reviews*, *44*(15), 5409–5445.
114. Young, T., Baka, J., He, Z., Bhattacharyya, S., & Lei, Z., (2023). Mining, loss, and despair: Exploring energy transitions and opioid use in an Appalachian coal community. *Energy Research and Social Science*, *99*, 103046.
115. Yu, J., Lucas, J. A., & Wall, T. F., (2007). Formation of the structure of chars during devolatilization of pulverized coal and its thermoproperties: A review. *Progress in Energy and Combustion Science*, *33*(2), 135–170.
116. Yuksek, O., Komurcu, M. I., Yuksel, I., & Kaygusuz, K., (2006). The role of hydropower in meeting Turkey's electric energy demand. *Energy Policy*, *34*(17), 3093–3103.

117. Zhengfu, B. I. A. N., Inyang, H. I., Daniels, J. L., Frank, O. T. T. O., & Struthers, S., (2010). Environmental issues from coal mining and their solutions. *Mining Science and Technology (China)*, *20*(2), 215–223.

118. Zhengfu, B. I. A. N., Inyang, H. I., Daniels, J. L., Frank, O. T. T. O., & Struthers, S., (2010). Environmental issues from coal mining and their solutions. *Mining Science and Technology (China)*, *20*(2), 215–223.

CHAPTER

2 An Organic Sediment

Contents

INTRODUCTION

Coal is an organic material from vegetative and animal partially decomposed materials. This process occurs in areas with saturated humid conditions, and the resulting product acts as an organic substance. This definition is relevant in both chemical-based and geological historical contexts (Terres, 1931; Lowry, 1945, 1963; Francis, 1961; Murchison & Westoll, 1968; Elliott, 1981; Bend et al., 1991; Bend, 1992). The circumstances mentioned above hindered the thorough decomposition of the residue (into the gases CO_2 and water) during its aggregation and subsequent transformation, ultimately resulting in the substance presently recognized as coal.

Sedimentary stones result from sedimentation procedures occurring on the outermost layers of the Earth, as in aquatic environments. Sedimentation encompasses a variety of mechanisms that lead to the settling and accumulation of minerals, while organic debris particles or just the precipitation of minerals from solution. The detritus comprised arboreal matter, papyrus-like vegetation, Lycopodiophyta (a taxonomic division of flora frequently categorized as the ferns along with club mosses), and numerous plant taxa that could have been discerned in coal deposits through their vestiges (Francis, 1961; Van Krevelen, 1961). Nonetheless, it seems that the creatures detected in numerous coal samples originated in environments with salty water.

Did You Know?

Coal is formed from ancient plants that lived and thrived during the carboniferous period, approximately 360 to 300 million years ago.

Following the sedimentation process, degeneration ensues, which refers to the suite of chemical, psychological, and biological alterations a sediment undergoes after its original deposition. Certain processes induce sediment consolidation, forming a dense, coherent substance from unconsolidated material. The process of sedimentation results in the interment of rocks, thereby causing a rise in the pressure and the temperature. This, in turn, triggers chemical reactions. One instance is the chemical transformations through which organic sediment (composed of organic matter) undergoes a coalification process, resulting in a coal formation. Upon an additional rise in either pressure or temperature, transformation succeeds in diagenesis, forming metamorphic material. In the context of the coal series with each other, anthracite is postulated to be a biological metamorphic rock.

Botanical remains of comparable nature are discovered within coal of varying ranks. As expected, due to the localized and regionalized distribution of flower species, there are likely to be substantial differences within the exact amount of such species across different sites (Stutzer, 1940). Furthermore, there exists the possibility of regional disparities in the environmental factors that facilitate physical development (as outlined in Table 2.1). These factors may encompass deviations in the amount of oxygen in the liquid, the pH balance, and the existence or non-existence of microbial organisms, alongside the deviations in flora types. The chemical composition of prenatal coal may be influenced by climate shifts among geological history eras/periods, which could also impact the different forms of plants (as shown in Table 2.2) (Bend et al., 1991; Bend, 1992; Speight, 2007).

Hence, it is unsurprising that coal exhibits significant variations in its composition across different geographical locations. It is not unusual to observe significant variations among the analytical characteristics of coal originating from a specific seam. This can be attributed to the diverse range of plant materials that might have served as the precursor and the multiple chemical processes that may have occurred during the maturation phase. The establishment of maturation indices has the potential to facilitate the identification of the specific path(s) along which growth transpired, as per scientific inquiry (Richardson & Holba, 1987; Ekwenchi et al., 1991).

Peat creation is a gradual process when plant debris accumulates under specific, albeit elusive, conditions. Despite not being officially designated as coal and typically excluded from the coal sequence, peat is widely considered the substance that emerges during the early phases of coal development.

Table 2.1. Representation of Variation in Conditions That Influence Coal Formation and Composition

Phase	**Variables**
The targeted degradation of carbohydrates along with lignin.	The variables of interest include water level, salinity, oxygen concentration, and temperature.
The level of waxy substances, cuticle resins, and microbes present in the sample.	The parameters of interest in this context are the pH level, which indicates the alkaline or acidic state of water, the oxidation and reduction possibility, and the duration of the measurement period.

The buildup of particulate matter consists of both inorganic and biological colloids.	The parameters of interest in this context are the pH level, which indicates the alkaline or acidic state of water, the oxidation and reduction possibility, and the duration of the measurement period.

Source: James G. Speight, Creative Commons License.

Table 2.2. Variations of Plant Forms with a Geological Period

Period	**Approximate Millions of Years Before Present**	**Botanical Characteristics**
Late Silurian	400	The initial occurrence of vegetation that has undergone lignification.
Carboniferous	270–350	The area boasts a vast and heterogeneous collection of spore-dispersing vegetation, encompassing ferns and svelte arboreal specimens exhibiting varying degrees of expanding and foliage elaboration.
Permian	225–270	The glossopteris, a type of seed-fern plant, exhibits widespread proliferation across the Gondwanaland region.
Triassic and Jurassic	180–225	The proliferation of plants that produce seeds is notable, particularly with the prevalence of conifers and cycadophytes.
Cretaceous	70–135	Angiosperms, commonly known as flowering plants, undergo evolutionary processes.
Late Cretaceous	~80	In most regards, the flora of today can be considered modern.
Tertiary	50–70	Grass species and sedges manifest.

Source: Heinz Heinemann, Creative Commons License.

Step in the process (Francis, 1961; Bouska, 1981), It is assumed that there is a sequential advancement within the coal series, wherein the lower-rank substances develop into higher-rank charcoal, as depicted in Table 2.3 and Figure 2.1. Peat is comparable to proto-petroleum, a poorly defined or unidentified substance believed as the initial product of organic matter's transformation into oil (Speight, 2007). There exist indications and theories suggesting that coal, along with terrestrial organic material of type III has the potential to serve as source material for petrol (Snowdon, 1991). The coalescence process is necessary for peat to transform into coal, as peat is a chemically intricate substance (Given, 1975). It is important to note that while peat moss is commonly believed to be the origin of coal, this theory has not been definitively established.

Geochemically speaking, coalification may be broken down into three separate procedures: (i) the micro-biological breakdown of the fiber in the original plant subject matter; (ii) the transformation of plant-derived lignin into humic compounds; and (iii) the condensing of those humic compounds into larger fossil fuels molecules. The type of decomposing vegetation, the state of decomposition, the surroundings in which deposits are made, and the motions of the earth's outermost layer are all significant determinants of the characteristics, excellence, & relative location of fossil fuel seams.

The procedure of coalification, which is influenced by local biological and geological factors, results in variations in the kind of coal developed. As such, the coal established is not uniform within a given country or geological epoch. Although some may argue that notable advancements have been achieved in this field, the verdict remains uncertain, resulting in extensive conjecture and the utilization of numerous methods for description.

Table 2.3. Description of Coal Types and Peat

Rank (Excluding Peat)	**Properties**
Peat	A conglomeration of recently amassed plant matter that has undergone partial carbonization and is not categorized as coal. The peat moss is a type of organic particle. The material exhibits a carbon concentration below 60% when measured on an empty ash-free foundation.
Lignite	Lignite, the least mature form of coal, may occasionally exhibit identifiable botanical features. As per its definition, the substance in question possesses a value for heating below 8,300 Btu/lb because it is devoid of mineral matter. When measured upon a dry ash-free foundation, the substance in question's carbon concentration ranges from 60% to 70%. Brown fossil fuels are used in Europe, Australia, and the United States to refer to certain types of low-level lignite for burning.

Subbituminous	The sub-bituminous material has undergone an elevated degree of organic metamorphism, resulting in the expulsion of certain proportions of oxygen and hydrogen from the coal. The depletion above results in the production of coal that exhibits elevated carbon concentration, ranging from 71% to 77% on an empty ash-free foundation. On a mineral matter-free basis, the heating value of sub-bituminous coal ranges from 8,300 to 13,000 Btu/lb. The classification of sub-bituminous coal through sub-bituminous A and sub-bituminous B or lower sub-bituminous C ranks is based on its value for heating.
Bituminous	Among the various ranks of coal, sticky coal is among the most prevalent in terms of its abundance. Approximately half of the coal generated in the coal generated to this factor. Bituminous coal is characterized by a percentage of carbon% to 87% on empty ash-free foundational issues, with significantly greater thermal efficiency than lignite along with sub-bitumen and Bitumen coals are categorized into Based on their volatile substance content, three subtypes, namely low-volatile bituminous, low-volatile-nous high-volatile bituminous, based on their volatile substance content. The term "soft coal" is commonly used to describe bitumen coal. It should be noted that this is a non-technical term and does not necessarily reflect the actual toughness of the coal.
Anthracite	Anthracite represents the utmost coalification and exhibits a carbon concentration exceeding 87% on an ash-free foundation. On an anatomically matter-free foundation, anthracite fuel coal has the greatest thermal efficiency per ton. The categorization of this substance is commonly based on its carbon content, leading to its division through anthracite and meta-anthracite. Anthracite is commonly known as hard coal. However, this terminology is colloquial and does not necessarily correspond to the coal's degree of hardness.

Source: Marcel Dekker, Creative Commons License.

Did You Know?

In ancient Rome, coal was primarily used as a fuel source for heating baths and buildings.

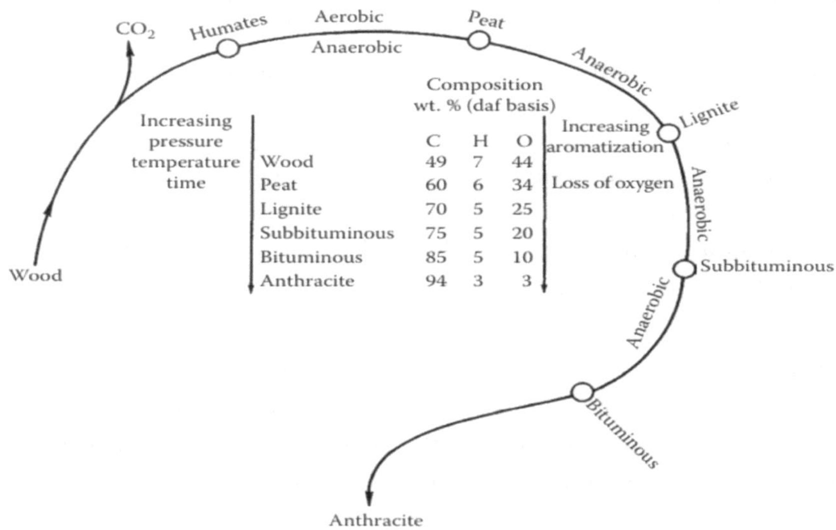

Figure 2.1. *Representation of coalification process.*

Source: John F. Fryer, Creative Commons License.

The correlation between the physical makeup of indigenous materials and the characteristics and real estate of the ultimate fossil fuels product still needs to be solved along with pending.

Since the 1980s, there has been an upsurge in coal scientific research; it needs to be clarified which ether there possesses been a commercialist advancement in comprehending color ocular composition and makeup of coal. The comprehension of the chemical-based molecular composition of coal presents a formidable obstacle that may remain unresolved indefinitely as coal.

Effective chemical precursors present in the initial mixture of pottage may form diverse hydrocarbons along with hetero atoms chemical groups with functional properties in coal. These indications may be identified through the application of different methods of analysis.

2.1. PRECURSORS OF COAL

The process of coal formation commenced throughout the period known as the Carboniferous Period, which is widely recognized as the initial coal age. Subsequently, significant coal reserves were established in each geological epoch, commencing from the Upper Carboniferous period approximately

270–350 billion years ago. The accumulation of plant material in damp or marshy areas with restricted contact with air and incomplete decomposition led to peat development. Through the process of burial by sediment, the peat moss underwent elevated temperatures and pressures, leading to a series of physical and chemical transformations that ultimately culminated in a coal formation. The formation of various coal ranks is attributed to the interplay of diagenetic and tectonic processes, which accompanied cycles of plant matter accumulation along with deposition. The extent of temperatures, time, and requirements exerted during these processes determined the ultimate coal rank.

The notion that coals originated through the organic remnants of plant matter is a well-established idea that emerged in the latter half of the 18th century, as documented by Moore (1940); and Hendricks (1945). The exact origins of this theory remain uncertain. However, identifying vegetative imprints inside coal deposits undoubtedly contributed to the notion that coal is closely linked to plant matter. Based on that proposition, it is a minor progression to the notion that plant matter is the natural antecedent to coal.

Undoubtedly, it is imperative to underscore the intricacy of coal and the coal-forming procedures, which necessitates the examination of plausible antecedents. Nonetheless, it is important to exercise prudence in assuming that minimal disparities will exist between contemporary organic substances and those present during the coal beds' deposition, except for the effect of natural selection. While this may seem like a plausible assumption, it is replete with unpredictability.

A comprehensive comprehension of the coal-forming mechanisms could be facilitated by obtaining a fundamental awareness of organic chemical compounds (Orheim, 1979). It is more advantageous to avoid attempting to comprehend the intricate and inadvertently distorted chemical processes frequently advocated as the outcome of a sequence of synthetic and possibly flawed circumstances of coal generation, along with the subsequent alterations in the chemical composition. It is important to bear in mind the characteristics of the hypothesized predecessors in the chemical makeup of coal.

Furthermore, despite inquiries regarding the chronology (both ordered along with concurrent) of the processes that took place throughout, it is imperative to utilize a suitable nomenclature. In vocabulary the context of this discourse, certain terminologies that are more expedient will be utilized. The expression "peat swamp," a multifocal and ecological entity described by (Given, 1975), refers to the early stages of anaerobic decomposition of woody substance. However, it should be noted that this is not intended to demonstrate conformity with the successive theory that posits that peat undergoes a gradual process of biological and physical changes to become coal. The reciprocal statement holds that using expedient, albeit outdated, nomenclature does not entail repudiation of a theoretical framework that espouses the principles of simultaneous chemical conversion sans the inclusion of coal varieties as intermediary substances in the progression Conversely, the phrase "organic debris" is utilized universally within this text to refer to any substance that has undergone destruction or deterioration yet has not undergone proper coalescence.

The composition of the organic constituents of plants encompasses sugars, cellulose, protein molecules, and various other polymeric materials. The organic nature of coal is commonly attributed to the presence of higher polymeric materials. However, it is important to note that the proportion of these components differs significantly depending on the specific species of plant and its stage of development.

Moreover, acknowledging that one among the above-mentioned compounds represents a wide-ranging classification (without considering specific molecular types) highlights the intricate nature of the initial plant matter. The complexity within the final fossil fuels molecule is primarily determined by this reality alone, which is likely the most significant contributing factor.

Notwithstanding, it is feasible (to simplify an intricate situation) to recognize certain generalizations; the lignin proportion in vegetation may vary from 10% to 35% w/w, whereas in the woody matter, the lignin may range from 25% to 30% w/w (each one of these estimates are expressed as ratios of the plant material's dry mass). The protein composition of grass ranges between 15% to 20% w/w, whereas woody materials exhibit negligible amounts of protein components. Additionally, grasses' fiber (carbohydrate) content falls throughout a range of 20% to 55% w/w. It is important to note that these ranges are approximations and are expressed as proportions of the weight at room temperature of the plant-based material).

Based on these overarching statements, it can be inferred that plant debris exhibits a high degree of complexity. Furthermore, the plant debris that gave rise to the antecedents of coal may become even more intricate, contingent upon the specific location and depositional events. Even the most proficient scientific investigator must acknowledge that the issue is significant because the complexity has increased over time.

The notion of a coal framework, commonly known as the average structure for coal, has persisted for numerous decades. However, its efficacy in advancing the understanding of coal remains a doubt among numerous engineers and scientists. It is unsure whether progress has been made in addressing the uncertainties surrounding coal in the four decades since its discovery. Some individuals possess the ability and inclination to present compelling arguments in support of either perspective on this matter. One may question whether it is necessary, or even refutable, to establish a precise molecular composition for coal.

The average makeup of coal is undermined by several factors, including the diverse chemical ingredients of coal, the varying geological environments in which it forms, the circumstances of development, and the variability of coal evaluation within a given seam.

Remember

Coal is not only used for energy production but also finds applications in the production of steel, cement, paper, and various industrial processes.

2.1.1. Lignin

Lignin (Sarkanen & Ludwig, 1971) is crucial in converting plant matter into coal (for discussion of this aspect of coal formation, please see Francis, 1961; Van Krevelen, 1961; Murchison & Westoll, 1968; Hyatsu et al., 1979, 1981; Lukoshko et al., 1979; Bredenberg et al., 1987; Hatcher & Lerch, 1991; Bend, 1992).

Although the molecular makeup of the lignin stays mainly hypothetical, some insights have been gained by identifying specific molecules present in the lignins of different kinds of plants. For instance, coniferyl alcoholic beverages (Formula 1; Diagram 2.2) were the primary constituent of conifer lignin (Masuku, 1992), whereas sinapyl alcoholic beverages (Formula 2; Diagram 3.3) are found in the fibers of deciduous plants and p-coumaric alcoholic beverages (Formula 3; Diagram 2.2) is present in the power source for gramineous (grasses).

Lignin is commonly recognized as a type of polymer and a combination of polymers composed of all the following alcohols. It has been postulated that identifying the number of individual alcohols can be inferred from assessing the power source methoxyl (–OCH) levels in diverse lignin varieties (Table

2.4). However, nitrogen within cellulose adds a layer of complexity to this process. The issue mentioned above remains unresolved in its entirety, and thus the origin of nitrogen in lignin remains uncertain. It is unclear how much nitrogen has become an inherent component of lignin, chemically bound within the lignin substance, or if it results from nitrogen-containing particles intricately intertwined with lignin and is challenging to isolate. However, it has been feasible to draw certain overarching conclusions regarding the characteristics of the connections present in this intricate organic compound. The carbon-carbon linkages present in the structure of the ring are classified as diphenyl kind (Formula 1; Diagram 2.3), while several either aliphatic carbon-carbon connections exist among them (Figures 2.2 and 2.3).

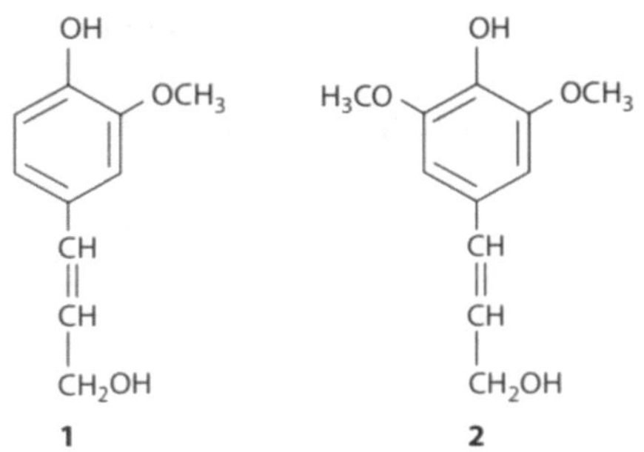

Figure 2.2. *Representation of monomeric units found in lignin.*

Source: Heinz Heinemann, Creative Commons License.

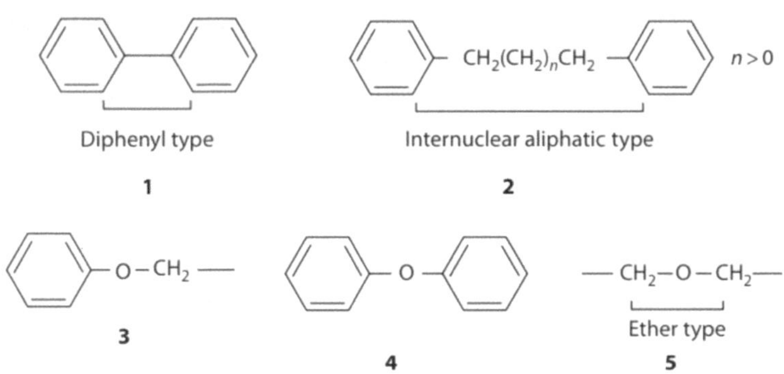

Figure 2.3. *Illustration of bond types that occur in lignin.*

Source: Marcel Dekker, Creative Commons License.

Table 2.4. Representation of Methoxyl and Nitrogen Content in Lignin from Various Sources

Lignin Source	OCH_3	N
Hardwoods	20.0	0
Softwoods	15.0	0.2–0.3
Graminaceans	10.0	1.2–1.6
Leguminoseans	5.0	2.9–3.4

Source: Murchison, Creative Commons License.

The theory proposing lignin as the primary organic precursor of coal lacks substantial empirical evidence. Extensive research has been conducted on the characteristics of lignin; however, there remain numerous aspects that have yet to be fully elucidated. Hence, the notion of an obscure chemical precursor (lignin) giving rise to, or being assimilated into, an enigmatic chemical derivative (coal) with purported comprehension of the exact chemistry may appear, and indeed is, somewhat odd from an academic standpoint.

2.1.2. Carbohydrates

Monosaccharides, also known as simple sugars, serve as the fundamental constituents of intricate carbohydrates and are ubiquitously present in nature, including prehistoric sediments (Swain, 1969). Polyhydric alcoholic beverages that contain between five and eight atoms of carbon are frequently referred to as sugars, with the five-along with six-carbon varieties being the most prevalent. Despite the apparent presence of a carbon dioxide (aldehyde and ketone) function within the basic formulas of sugars, this feature is not manifested. Typically, the carbon-containing moiety is found in conjunction with any of the hydroxyl groups present within the same molecular entity, forming a periodic hemiacetal or hemiketal.

To the extent that monounsaturated are present in nature, they are more prevalent to observe their natural occurrence in pairs (disaccharides) alongside threes (trisaccharide) and more frequently as high-molecular-weight polymers. The source material is likely to be primarily composed of polymers, with particular emphasis on two widely recognized polysaccharides, namely starch. The plant and tree cell wall's sticky substance comprises cellulose and starch and is present in diverse forms throughout the vegetative kingdom, typically serving as a nutritional reserve. The chemical constitution of starch varies depending on its origin; however, within a single starch sample, two distinct polymers with different structures can be identified. Typically, amylose and amylopectin are composed solely of glucose units; however, amylose is characterized by a linear structure, while amylopectin exhibits a branched structure.

Furthermore, compared to the relatively conjectural chemical reactions of lignin, the basic chemical reactions of numerous polysaccharides, particularly starch, and cellulose, are typically better established (as illustrated in Figure 2.4). The fiber's molecular structure is $(CHO_5)_n$, where the value of n can reach a few 1,000. The hydrolysis of cellulose using strong hydrochloric acid as a solvent result in a significant yield of glucose, typically ranging from 95% to 96%. This suggests that glucose is the primary monomeric unit present in cellulose.

Therefore, a plausible molecular configuration for cellulose is founded upon the stereochemical organization of glucose (as depicted in Formula 1 and Figure 2.5). Furthermore, it should be noted that the cellulose molecule exhibits a non-planar conformation featuring a "screw axis" whereby each glucose unit is positioned at approximately 90° to the subsequent unit. While it may seem plausible that free rotation could occur around the –C–O–C– bond, it is important to consider the influence of steric effects, which ultimately restrict the degree of rotation that can take place. Moreover, the compact arrangement of atoms, coupled with restricted rotation, results in a stiff molecule. Additionally, the intermolecular interaction of hydrogen that binds the lengthy chains together confers a three-dimensional shape upon cellulose.

Although the current depiction of cellulose is simplistic, it does not encompass the entirety of the molecule's structure. Cellulose exhibits a low frequency of achieving complete glucose yields upon hydrolysis. Additionally, there has been speculation regarding the potential presence of non-carbohydrate groups as a source of origin. In brief, it should be pointed out that the chemical makeup of cellulose is considerably more intricate than the simplified representation provided herein. Nonetheless, as mentioned above, the structure is sufficient for elucidating cellulose's fundamental chemistry principles. It is worth noting that the clear framework of coal may not be as well-defined when utilizing the results of destruction experiments to define its structure. This presents a peculiarity.

Due to its polysaccharide nature, the chemical properties of starch (as depicted in Figures 2.6 via 2.8) are frequently regarded as analogous to those of cellulose. While this could not be entirely applicable to proficient biochemical researchers, it is satisfactory for the current discourse, which aims to demonstrate the intricacy of the molecular components hypothesized to have played a role in coal formation by contributing to plant material (Francis, 1961; Van Krevelen, 1961).

In broad terms, it can be stated that plant debris may have contained two polymers with a high molecular weight, namely starch along with cellulose, which has a broad chemical formula of $(C_6H_{10}O_5)_n$. The value of n could range from a few 100 to a few million units. These polysaccharides transformed and became part of the coal framework.

2.1.3. Proteins

According to Hare (1969), amino acids are organic compounds that include nitrogen and are present within the protoplasm of cells in both animals and plants. Protein structure is subject to variation depending on the source. However, it is possible to infer an approximate range for the protein makeup (as presented in Table 2.5). Additionally, certain proteins contain trace quantities of additional elements, such as phosphorus (in nucleoproteins) and iron (in hemoglobin).

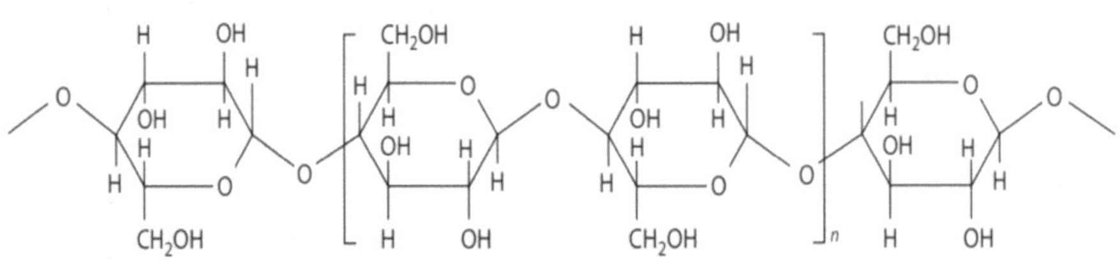

Figure 2.4. *Illustration of the molecular structure of cellulose.*

Source: James G. Speight, Creative Commons License.

Proteins primarily comprise a chain or ring of amino acid sequences interconnected by amide linkages, as depicted in Figure 2.9. Peptides and proteins belong to a comparable category of chemical compounds, although the differentiation between the two is not always unequivocal. A common method for distinguishing among proteins and peptides is to utilize a molecular weight threshold of 10,000, whereby proteins are defined as having an atomic weight greater than 10,000. In contrast, amino acids possess a molecular mass below this threshold. The differentiation could also be established based on variances in physical characteristics, such as the level of water intake and spatial configuration of the molecule. As an illustration, naturally occurring amino acids possess comparatively brief and pliable chains and exhibit the potential for reversible dehydration despite being hydrated in a water-based solution. In contrast, proteins consist of lengthy chains that exhibit distinct conformational properties, with interstices being occupied by water molecules.

The unique attributes of proteins are attributed not solely to their extended and intricate polypeptide chains but also to how organisms biosynthesize them in a frequently unstable state that can be readily modified (denatured). Incidentally, boiling an egg can illustrate protein denaturation, wherein the egg albumin undergoes such changes conveniently. Furthermore, the spatial conformation of proteins is characterized by a three-dimensional structure, which exhibits a fascinating setup that includes an α-helix (Figures 2.5–2.7).

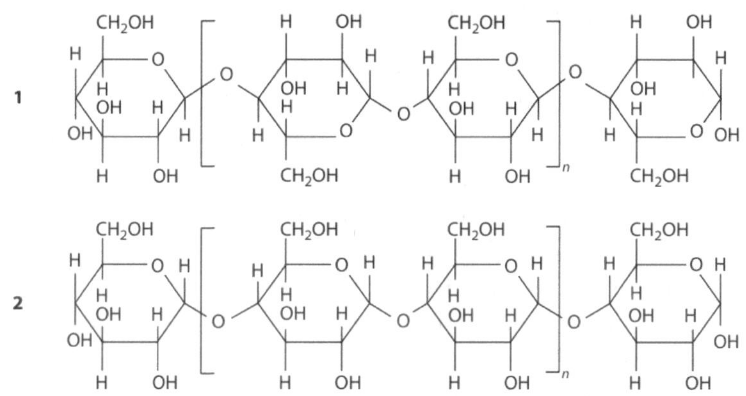

Figure 2.5. *Structures of cellulose and starch.*

Source: John F. Fryer, Creative Commons License.

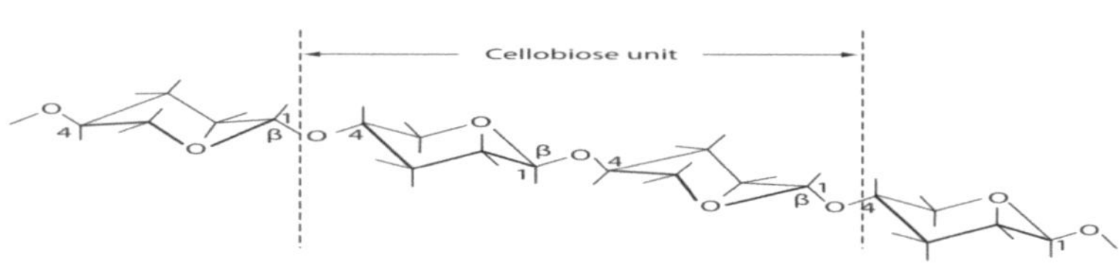

Figure 2.6. *3D representation of cellulose.*

Source: Francis, Creative Commons License.

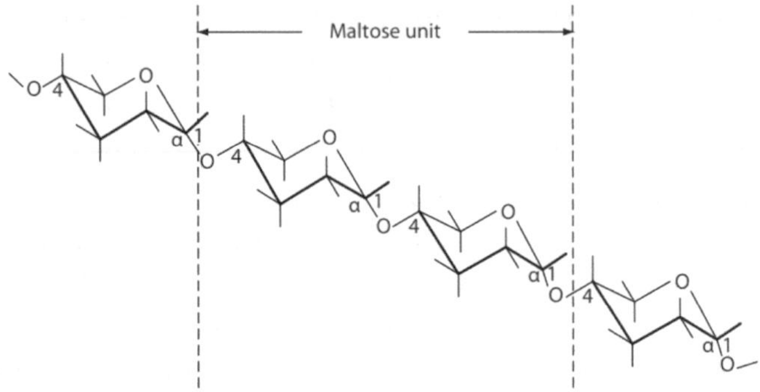

Figure 2.7. *3D representation of starch.*

Source: Van Krevelen, Creative Commons License.

The salient characteristic of this helix pertains to the arrangement of peptide chains, which coil in a manner that facilitates the establishment of hydrogen bonds among the amides hydrogen fuel atom(s) along with carbonyl group(s) at strategically positioned locations along the chain. The formation of said bonds can be attributed to the polarization of diverse organic functional groups, wherein the hydrogen atoms in close proximity exhibit a positive division, and a heteroatom (typically oxygen) displays a negative polarization. The magnitudes of the advantages of such bonds are typically in the range of 5–10 kcal (2.1–4.2 kJ), which is notably lower than the advantages of traditional natural σ-bonds, which can range from 60–100 calories (25.1–41.9 kJ) (Figures 2.8 and 2.9).

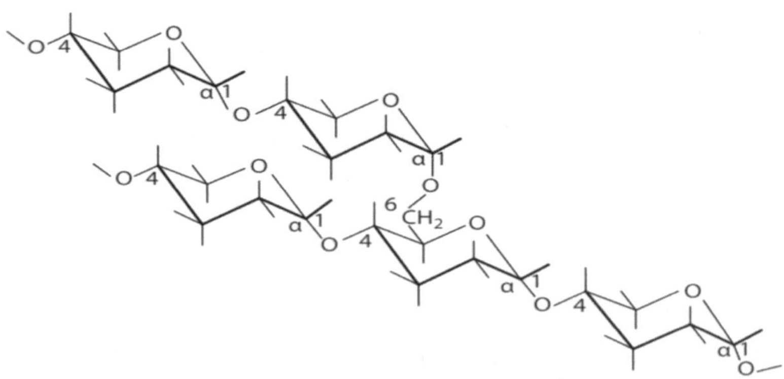

Figure 2.8. *3D representation of starch (amylopectin).*

Source: Francis, Creative Commons License.

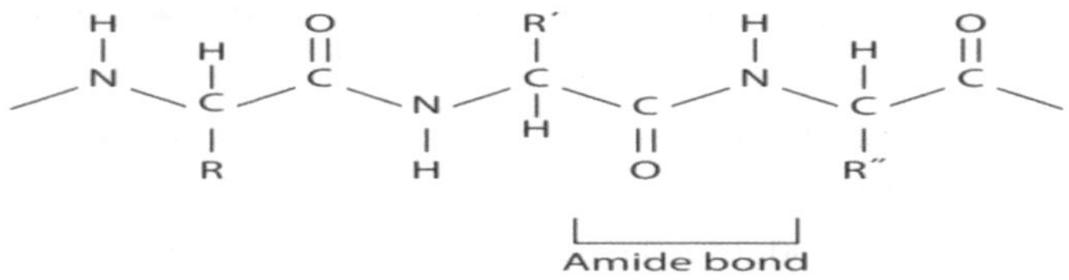

Figure 2.9. *Illustration of part of the protein.*

Source: Van Krevelen, Creative Commons License.

Table 2.5. Range of Protein Composition

Component	Percent by Weight
Carbon	47–50
Hydrogen	6–7
Nitrogen	15–18
Oxygen	23–26
Sulfur	<1

Source: James G. Speight, Creative Commons License.

The protein's double helix formation can form extremely stable hydrogen connections, leading to ring systems that provide extra stability to the whole thing, typically ranging from 10–20 calories (4.2–8.4 kJ).

The lateral chains of amino acids are situated externally to the helical coil. Notably, the perfection of α-helices within proteins is not universal, as the rigidity of the spiral may be affected by steric effects arising from specific chains of amino acids.

Proteins are ubiquitous and exhibit a diverse range of sizes and configurations. Many proteins incorporate metallic elements, including iron, zinc, and copper, which play a crucial role in the physiological processes of the associated molecules.

2.1.4. Oils, Fats, and Waxes

Plant-based oils, fats, along with waxes are frequently encountered components and typically comprise terpene compounds, hydrocarbons along with their derivatives, and alcoholic beverages, acids, and ester compounds (Parker, 1969; Streibl & Herout, 1969; Wollrab & Streibl, 1969).

Nevertheless, oils are present in significantly higher quantities in seeds and fruit. Organic substances, commonly referred to as essential oils, are characterized by their pleasant aroma and may comprise sulfur, nitrogen, and carbon, or hydrogen. The oils frequently belong to the terpene category and exhibit a general formula of $(C_5H_8)_n$, where n is greater than one. Alternatively, they may be compound varieties that bear a resemblance to gasoline or additional products of oxidation that are linked to the fundamental terpene structure.

Alternatively, the oils present in plants can belong to the categories of fatty acid or glyceride, or they may also exist as esters that contain more monohydric spirits fatty acids. Glycerides, which are esters, are the prevalent form in which acids are present in plants.

$C_{15}H_{31}COOH$	$C_{17}H_{35}COOH$	$C_{17}H_{33}COOH$
Palmitic acid	Stearic acid	Oleic acid

Symmetrical geometry glycerides can be produced when glycerol combines with any of the preceding fats in which their acid-free radicals are identical. This is observed within glycerides:

Tripalmitin	Tristearin
$CH_2OCOC_{15}H_{31}$	$CH_2OCOC_{17}H_{35}$
$CHOCOC_{15}H_{31}$	$CHOCOC_{17}H_{35}$
$CH_2OCOC_{15}H_{31}$	$CH_2OCOC_{17}H_{35}$

Various types of fatty acids can bind with a single molecule regarding glycerin, forming a heterogeneous glyceride.

Palmitodistearin	Oleodistearin
$CH_2OCOC_{17}H_{35}$	$CH_2OCOC_{17}H_{35}$
$CHOCOC_{17}H_{35}$	$CHOCOC_{17}H_{35}$
$CH_2OCOC_{15}H_{31}$	$CH_2OCOC_{15}H_{33}$

Vegetable oils, as well as fats, typically exhibit a liquid or semisolid state at room temperature along with may lack a clearly defined point of melting. A significant impact on human health. Furthermore, incorporating the substance, as mentioned earlier, into a coal precursor has increased complexity regarding the involvement of fatty acids. Sterols, including chemical compounds, are present in vegetative and animal oils and fats. The configuration of sterols is founded upon the poly-cyclic perhydrocyclopentaphenanthrene framework, as depicted in Figure 2.10. Sterols are characterized by their crystalline nature and the presence of a hydroxyl group, as exemplified by cholesterol. These compounds can be found in either their uncombined state or as fatty acids containing more fatty acids. They can be extracted from a nonsaponifiable fraction of fatty acids and oils. Sterols and their corresponding substances, steroids, are also present in animal tissue. In such instances, it is plausible that the substance in question could be cholesterol, or modifications may have been made to either the hydroxyl group of the molecule or the alkyl group opposite chain, producing a structurally analogous compound. In the latter scenario, it is postulated that the predominant source of the steroid-like material that facilitated coal formation may not have originated from animal remnants. Rather, it is suggested that a great deal of the power source steroid-B-type evidence that ultimately constituted the resulting coal substance was derived from plant detritus.

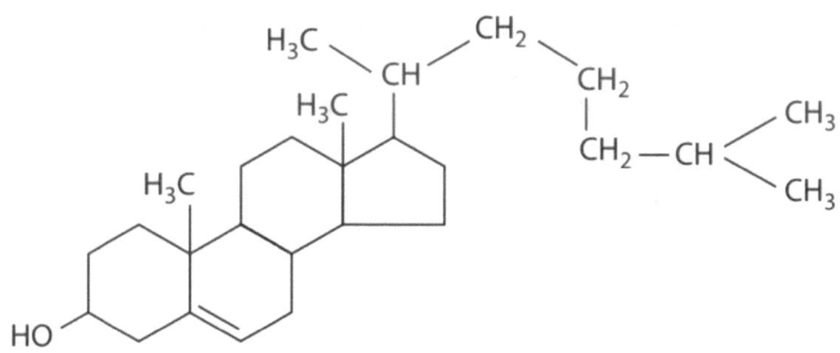

Figure 2.10. *Representation of cholesterol showing four-ring steroid nucleus.*

Source: Marcel Dekker, Creative Commons License.

Waxes are frequently encountered naturally as alcoholic beverages and esters alongside high molecular weights. However, it is also possible for free fatty acids to be included within this category. Natural waxes possess a distinct constitution compared to crude oil wax, solid at ambient temperature and composed of straight-chain paraffin wax petroleum products with a carbon atom count exceeding 17. Natural waxes lack functional oxygen groups (Speight, 2007).

The molecular chemistry of organic waxes is comparatively less clear than that of saturated fats and oils (glycerides). Although the nomenclature of straight-chain content, such as octacosanol, a 28-carbon alcoholic beverage, is well-established, there remains uncertainty regarding the structural composition of certain natural waxes named less conventionally. For a considerable period, opinions have been divergent regarding the precise configurations of certain compounds, such as montane alcohol, that conceivably represent a blend of alcohols exhibiting distinct carbon numbers. Consequently, the carbon count of several waxes that possess comparatively vague designations may be legitimately disputed.

2.1.5. Miscellaneous Organic Materials

2.1.5.1. Resins

Resins are a class of substances that are released from injuries in the outermost layer of various species of trees. Various forms of resins are present, and each variant seems to possess distinct attributes specific to the category of trees that produce them.

Due to the vast number of possible substances that may be present in a resin combination, it proves to be a challenging task to attribute a definitive structure to any given resin. It is widely acknowledged that the resin colophony, also called rosin, is a nonvolatile residue obtained from the boiling process of turpentine. This residue is composed of a blend of resin acidic substances from the diterpenes. Abietic acid, a tricyclic diterpene (as shown in Figure 2.11), is a well-known component of this particular mixture).

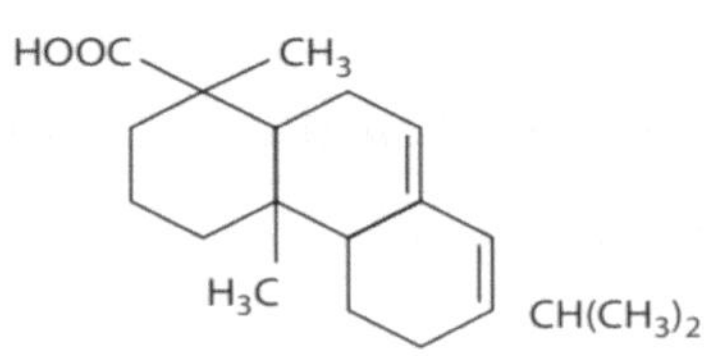

Figure 2.11. *Illustration of abietic acid.*

Source: Meinschein, Creative Commons License.

2.1.5.2. Alkaloids

Alkaloids represent the final category of organic substances derived from plants that hold significance within coal's organic matter. The molecules for these substances possess basic nitrogen, often in a cyclic structure. Furthermore, a majority of the alkaline substances additionally comprise oxygen. Alkaloids are compounds in various plant parts such as roots, tree bark, branches, and intracellularly (Dalton, 1979).

The variability among the structural chemical reactions of alkaloids can be attributed to the diverse locations where nitrogen may exist within organic systems. It is commonly acknowledged that alkaloid compounds can be derived from various distinct systems, individually or through an amalgam of multiple systems.

2.1.5.3. Porphyrins

The pyrrole powder system gives rise to another category of compounds made up of nitrogen, namely the porphyrins, as illustrated in Figure 2.12. The porphyrins are derived from a conjugated cyclic framework known as porphine, which comprises four pyrrole starts ringing interconnected via methine (=CH–) crossings at the 2 along with 5 places (*refer* Figure 2.12). While porphine is not naturally occurring, its derived structures play a crucial role in the survival of both flora and fauna (Baker, 1969). The phosphine system is present in the vertebrate kingdom as hemoglobin, specifically in the power source hemin component of the molecule. This system is an oxygen transporter to the lungs and body tissues via the red blood cells. Porphyrin is present within the chlorophyll in the stems and leaves of plants, crucial for photosynthesis in the plant kingdom. Bonnett et al. (1991) have suggested that the porphyrin components present in coal could serve as useful indicators for assessing the degree of coalification, much like how porphyrin components in petroleum are utilized to track the progression of petroleum development (Branthaver, 1990).

2.1.5.4. Hydrocarbons

The polyenes represent a noteworthy class of substances that may contribute to the formation of coal precursors. The substances above include bicyclic alkaline compounds, aromatic compounds, hydrocarbons, and pigments called carotenoids (Meinschein, 1969; Schwendinger, 1969; Luo et al., 1991).

Furthermore, the presence of picene derivatives and analogous polynuclear fragrant systems (White, 1983) in solvent-based coal extracts can be considered as substantiation over sterane-type substances in the antecedent matter of coal. This presupposes preserving the skeletal framework during the developmental phase without any deterioration or emergence of novel skeletal structures. Producing aromatic substances through biosynthesis has been extensively documented, encompassing the conversion of nonaromatic predecessors into aromatic organisms (Weiss & Edwards, 1980; Lee et al., 1981).

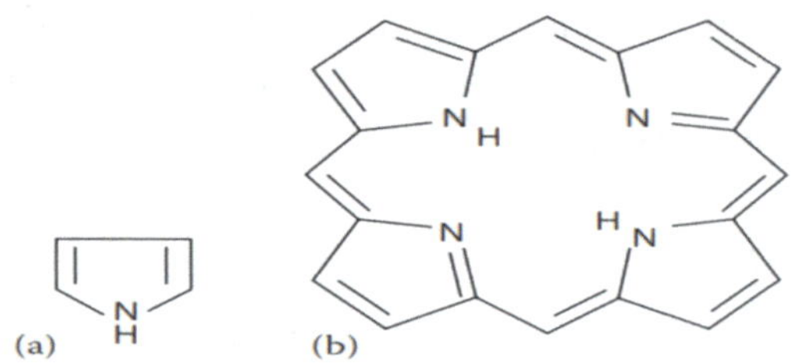

Figure 2.12. *Illustration of (a) pyrrole, the major building block of porphyrins; and (b) porphine, the major structural unit of porphyrins.*

Source: Branthaver, Creative Commons License.

2.1.5.5. Inorganic Constituents (Mineral Constituents) of Plants

The preceding segments have provided comprehensive coverage of the plant's organic components that may serve as the origin material. An additional category of chemical compounds is derived from plants and subsequently incorporated into the lignite matrix. This is a rock-like material that originates from the inorganic components of plants.

Coal's mineral makeup is typically categorized as related mineral matter or extraneous minerals. The inorganic substance in the originating bed is inherent mineral content, which originates in the plant substance that constitutes the organic debris. Conversely, extraneous mineral matter refers to the inorganic substances introduced within the coal-forming release through diverse mechanisms from external origins.

While it is true that a great deal of the minerals found in most coals is derived from external sources, it remains important to acknowledge the inorganic components of the original plant materials. The extraneous inorganic matter that ultimately integrates into coal may have diverse origins. Phosphorus is present in amino acids, magnesium ions are in chlorophyll, and several plants contain sodium, calcium, iron, and potassium. Mineral content is a ubiquitous component of plant cell exteriors and may potentially form a chemical bond that links the organic material of plants. Presently, it is widely accepted that the predominant source of nickel alongside vanadium is caused in petroleum emanating from the assimilation of the chlorophyll (in addition to chlorophyll-type) substances into the origin stratum. These substances are inherently abundant in the metals as mentioned above, present in a chemically bonded state with the organic component. Similar observations could apply to analogous constituents present in coal.

Typically, the quantity of inorganic matter in the woody parts of trees is below 1% w/w. Conversely, the leaf tissue of trees may contain approximately 2% to 3% w/w of inorganic material, with the outside tissues (cork) of plants containing upwards of 20% w/w of organic matter. Iron is primarily present in the leaves at an individual level. At the same time, the sap comprises a mixture of sodium, potassium, and magnesium oxide, along with calcium solutions, along with a certain amount of iron.

While the overall percentage of innate inorganic material in coal reserves is typically low, certain constituents of coal might have a relatively significant amount of innate inorganic matter. Coal components that exhibit elevated cuticle levels can also show heightened levels of intrinsic iron compounds compared to coal components that originate from alternative plant structural elements.

2.2. HETEROATOMS IN COAL

Coal is applied in various processes, including combustion, carbonation, melting, gasification, and manufacturing chemicals. The varied composition of coal is determined by its origin, necessitating the characterization of coal before its utilization, whether or not as singular or composite coal.

Characterizing coal to ascertain its properties is employed to anticipate its technological performance. The utilization of coal is generally influenced by two key factors: its composition and rank. The chemical makeup of coal is characterized by two factors: the type that pertains to the natural components, their nature the grade, which can be ascertained by the number of mineral deposits in the coal. The grade measures the degree to whereby the mineral content dilutes the matrix of organic substances.

It is crucial to examine the hetero atoms organisms present in coal, which refer to the non-hydrocarbon species containing nitrogen, oxygen, and sulfur. These species can originate from plant predecessors or external sources that may arise throughout maturation. These species play a significant role in determining the magnitude of airborne emissions, which can harm coal utilization as an energy source and may elicit negative feedback.

2.2.1. Nitrogen

Despite being one of the principal constituents of coal, the chemical compound nitrogen has yet to receive much attention in research. However, it is noteworthy that the nitrogen content in coal is typically incorporated into its organic carbon-based component, with few exceptions. Limited data is accessible regarding the nitrogenous substances in coal; however, they seem steadfast and are believed to be predominantly heterocyclic. It is plausible that the primary origin of nitrogen present in coal deposits could have been derived from the proteinaceous matter of both flora and fauna. Nitrogen within coal can be attributed to the stable cyclical structures in plant alkaloids, chlorophyll, and various other porphyrins, which have endured the coalification process. There appears to be a weak correlation between coal rank and the quantity of nitrogen present in coal (Davidson, 1994).

Coal usually contains nitrogen within the range of 0.5% to 1.5% by weight. Similar to oxygen, various compounds with nitrogen in them have been extracted from liquids derived from coal, as reported by Schiller in 1977. Various compounds fall under this category, such as isoquinolines, quinolines, pyridine, carbazoles, and anilines, along with indoles, which possess alkyl and aryl substitutes (Schweighardt et al., 1977; Palmer et al., 1992, 1994).

The quantitative determination of five significant nitrogen chemical structures found in coals of different ranks has been achieved by applying nitrogen X-ray consumption near-edge spectrum (XANES). For the past decade, there have been investigations into the sulfuric acid chemical makeup of coals. However, it is only in recent times that analyses of nitrogen in these petroleum specimens have been done. The X-ray absorption near edge structure (XANES) spectrum of coal samples displays multiple discernible peaks that can be associated with specific oxygen chemical structures, enabling the examination of these intricate systems. The XANES spectrum of coals was comprehensively analyzed to account for all features, leading to the identification of the major structural joblessness groups of nitrogen identified in coals. The coals contain diverse aromatic nitrogen-containing substances, while no indication of soaked amino acids was detected. Coal contains various derivatives such as the pyrrole pyridine along with aromatic amines. Among these, pyrrole constructions are the most abundant. The presence of nitrogen within pyridine is observed in all coal types except for those classified as low-rank. The high pyridone content in low-rank coals alongside high oxygen levels is associated with a low pyridine compound content. As mentioned earlier, the observation implies that the process of coal maturing leads to the conversion of pyridone into pyridine, as the former loses its oxygen content. Aromatic amines are detected in trace amounts in coals of varying degrees of maturation (Mullins et al., 1993).

The potential correlation between nitrogen oxide (NOx) that is released resulting from coal combustion and the nitrogen composition of the coal has been explored in the literature. It has been observed that a greater proportion of nitrogen may be retained in the char (Hindmarsh et al., 1994; Wang et al., 1994) residue as opposed to petroleum-based coke (Speight, 2007).

The relationship between coal properties and NOx emission levels needs a clear definition. The formation of NOx is influenced by multiple factors, including but not limited to nitrogen dividing among volatiles and chars, nitrogen performance, coal position, coal amount of nitrogen, and volatile content. These factors play distinct roles and interact with each other in NOx formation. Numerous studies have consistently shown that utilizing coal nitrogen level as a criterion for coal selection is an unsuitable and inefficacious approach for limiting NOx emissions. The discretion to choose coals ought to be exclusively vested in the generating utility, which possesses the most comprehensive comprehension of the impact of their combustion plant's layout and operation on the magnitude of emissions of NOx. In addition, the persistent utilization of fossil fuels nitrogen as a regulatory criterion for coal provisions to a burgeoning utility sector will not solely prove ineffective in constraining NOx discharge levels but also impede the advancement of novel coal reserves.

Although the impact of nitrogen function upon the release of NOx remains uncertain, several researchers have discovered that pyrrole-type nitrogen buildings are notably less stable than pyridine-type structures (Davidson, 1994). Consequently, pyrrole-type things are more susceptible to oxidation and reaction, leading to the formation of NOx. This factor was observed to affect NOx production through both char and volatile burning. When considered in isolation, the nitrogen level of coal does not indicate the exact proportions of distinct nitrogen groups with functions present in a given coal sample (Keleher, 1995).

2.2.2. Oxygen

Oxygen is present in every one of coal's inorganic and organic components. Within the organic component, the presence of oxygen is typically observed in hydroxyl (–OH), alcohol groups, carboxyl chains (CO_2H), methoxyl categories ($-OCH_3$), as well as groups of carbons (=C=O). Low-rank coal typically exhibits a hydroxyl oxygen content ranging from 6% to 9%, whereas high-rank coal generally contains less than 1% hydroxyl oxygen content. The oxygen percentages in carbonyl, methoxyl, and carboxyl compounds exhibit an average range of a few percentage points in low-rank and brown coal. In contrast, in high-rank coal, the values are almost negligible and cannot be measured accurately. Coal comprises inorganic constituents that contain oxygen, including but not limited to moisture, silicon dioxide, calcium carbonate, oxides of iron, and sulfur compounds. The predominant constituents of the shale-like parts are metal silicates, commonly called silicates. The predominant constituent of carbonate is a carbonate of calcium ($CaCO_3$), while the oxides are primarily composed of iron oxides (FeO alongside Fe_2O_3), or the sulfates consist of iron and calcium ($CaSO_4$ and $FeSO_4$).

Oxygen in coal is believed to originate from various plant sources. However, it is also highly probable that the addition of oxygen occurs throughout the maturation process by exposing the precursors to atmospheric oxygen and oxygenated waters that infiltrate the detrital payments. The oxygen weight percentage weight in coal varies depending on its type, ranging from 20% to 30% for lignite mining and approximately 1.5% to 2.5% for anthrax.

Typically, the oxygen content in a specific coal is determined through different methods. The equation for determining the percentage of oxygen in a substance is 100 minus the sum of the percentages of carbon, hydrogen, nitrogen, sulfur, and ash present in the substance (all measurements were conducted on a dry weight basis) %Oxygen = 100 – (%C + %H + %N + %S + %ash) (all calculated on a dry basis) Coal is recognized to contain diverse forms of oxygen, such as phenolic hydroxyl, amino carboxylic acid, carbonyl, and ether connections, along with heterocyclic oxygen, through respiration.

The rate of coalification significantly influences the occurrence of the acids carboxylic and phenolic hydroxyl. The total volume of both groups exhibits a significant decrease to be the carbon content (rank) increases. Peat and lignite can facilitate the extraction of compounds that comprise the carboxylic acid moiety (–CO_2H). However, these compounds appear to be absent within the sub-bituminous range. Similarly, the methoxyl grouping (–OCH_3) has been observed in lignite. However, its presence in bitumen and higher-rank coal is yet to be shown to be substantial (Deno et al., 1978).

The presence of phenolic substances hydroxyl, ether connections, and heterocyclic oxygen has been observed in bituminous coals, as well as in higher-rank coals, to a lesser degree. The study reveals a gradual reduction in the phenolic hydroxyl material with increased carbon.

Furan ring structures are the prevailing form of heterocyclic oxygen in the air. The presence of furan rings that have been replaced with other chemical groups has been documented in various sources, such as coal extracts, burning tars, and byproducts resulting from oxidation. The precise extent of the presence of the dibenzofuran system in the original coal source remains uncertain. Identifying dibenzofuran within oxidative products (Duty et al., 1980; Bodzek & Marzek, 1981; White, 1983) of decomposition and low-temperature coal extracts proves its existence in the original coal sample.

2.2.3. Sulfur

The origin of sulfuric acid in coal needs to be better understood and is believed to result mainly from containing sulfur plant parts such as oils and proteins. However, there are alternative mechanisms through which sulfur may be incorporated into coal. This has been discussed in several studies. How sulfur is incorporated into coal is primarily determined by the characteristics of the coal.

The significance of sulfuric acid in coal usage has led to a substantial body of literature focused on enhancing the efficacy of techniques and refining the precision and the precision of sulfuric acid measurement (Speight, 2005 and the sources cited therein).

Coal contains sulfur in three distinct forms, namely organically bound sulfur, inorganic sulfur in the shape of pyrite and marcasite (FeS_2), and inorganic sulfates. This information is supported by various sources, including ASTM (2011); ISO (2011); Wawrzynkiewicz & Sablik (2002); and Speight (2005), along with their respective references. Typically, the coal's organic sulfur content is below 3% w/w. However, there have been instances where the sulfur content has been extremely high, reaching up to 11%. The presence of sulfates, specifically calcium sulfate ($CaSO_4$) and iron sulfate ($FeSO_4$), is typically limited to a maximum of 0.1% in coal samples unless they have undergone significant weathering or oxidation. The two prevalent crystal structures of FeS_2, namely pyrite, and marcasite, pose a challenge in differentiation and are frequently misidentified as pyrite.

It has been observed that certain coal samples exhibit a rise in overall sulfur content in the presence of the two marine invasions and organic matter. The distribution of sulfur is contingent upon the organic composition. In clay-rich environments, most sulfur exists in the form of pyrite. Conversely, in samples with high levels of organic matter, the sulfur content is typically dominated by organic sulfur, which is present in concentrations roughly equivalent to pyritic sulfuric acid (Bailey et al., 1990).

Coal does not contain a significant amount of free sulfur. There is significant variation in the quantity of sulfur-containing substances in coal, particularly among coals sourced from distinct seams. Furthermore, the distribution of pyrite in coal is not homogeneous. It can manifest as strata or plates or may be dispersed throughout the organic matter as minuscule crystals. The concentration of sulfates, predominantly consisting of Gypsum out ($CaSO_4 \cdot 7H_2O$) as well as ferrous sulfates ($FeSO_4 \cdot 7H_2O$), is typically minimal (i.e., $<0.1\%$) in coals unless they have undergone significant weathering or oxidation.

During coal mining, newly discovered minerals containing sulfur within the coal and surrounding rocks get exposed to the elements of air and water. The chemical reactions that occur yield sulfuric acid along with precipitates. Untreated acid water emanating from coal mines has the potential to cause harm to the biota present in the streams that receive it. The deposition of iron and sulfate frequently results in the chromatic alteration of stream beds, manifesting as yellow and orange discolorations. Likewise, upon combustion, sulfur present in the exhaust fumes has the potential to react with atmospheric water, resulting in the formation of acidic precipitation, commonly known as acid rain. Using high-sulfur coal within power plants can result in corrosion of the metallic apparatus due to similar reasons.

According to Casagrande et al. (1979), the origin of organic sulfur in certain coals can be attributed to the reaction between peat-like precursors and hydrogen sulfide, accounting for more than half of the entire sulfur content. Conversely, multiple strands of proof (Chou, 1990) suggest that the primary origin of sulfur within low-sulfur coal is plant material, while in medium-along with high-sulfur coal, ocean water is also a significant contributor of sulfur (Chou, 1990).

2.2.4. Other Elements

The occurrence and quantity of diverse constituents in coal are contingent upon the origin of the coal, as stated by Speight (2005) and the sources mentioned therein.

Chlorine is present in coal and is speculated to contribute to fouling issues and corrosion-related concerns. The coal's chlorine concentration is minimal, often comprising merely a fraction of a percent or less. The primary constituents of this substance are the minerals calcium, sodium, and potassium chlorine, while certain coal samples may also contain iron and magnesium chlorine. There exists evidence indicating that chlorine has the potential to react with the organic constituents present in coal.

Mercury has been recognized as a hazardous environmental pollutant primarily due to its tendency to accumulate in the food chain through a concentration procedure. The existence of mercury within coal is a highly delicate matter. The potential release of mercury that could be present in coal is a matter of environmental apprehension.

According to Tewalt et al. (2001), a sequence of chemical reactions can lead to converting elemental mercury, which is already a toxic substance, into a significantly more hazardous form. Methyl mercury is widely regarded as the most hazardous form due to its organic nature, produced through a sophisticated bacterial transformation of inorganic mercury. This variant has the propensity to infiltrate and amass within the food web, predominantly in aquatic creatures like fish and birds, leading to a range of illnesses in both animals and humans. The production of methylmercury within ecosystems is contingent upon several factors, including the abundance of the element mercury, bacterial population structure, nutrient levels, acidity, and oxidation conditions, amounts of sediment, and sedimentation process rates. According to Tewalt et al. (2001), a sequence of chemical reactions can lead to converting elemental mercury, which is already a toxic substance, into a significantly more hazardous form. Methyl mercury is widely regarded as the most hazardous form due to its organic nature, produced through a sophisticated bacterial transformation of inorganic mercury. This variant has the propensity to infiltrate and amass within the food web, predominantly in aquatic creatures like fish and birds, leading to various illnesses in humans and animals. The production of methylmercury within ecosystems is contingent upon several factors, including the abundance of the element mercury, infection population density, nutrient levels, acidity, and oxidation conditions, amounts of sediment, and sedimentation rates over time.

During the process of burning coal, the predominant form in which mercury escapes is elemental mercury, owing to the thermodynamic equilibrium that favors this condition at the elevated temperatures of coal combustion (Lu et al., 2007). Elemental mercury (Hg) is vaporized through the coal as it undergoes combustion. The oxidation of elemental mercury to produce mercuric oxide (HgO), mercuric chlorides ($HgCl_2$), along with mercurous chloride (Hg_2Cl_2) is facilitated by the amount of chlorine in the system as the temperature of the flue gas declines.

Most coal deposits typically comprise trace quantities of mineral carbonates, predominantly composed of magnesium carbonate and, to a lesser degree, ferrous as well as other metal carbonate. Certain coal samples exhibit a relatively substantial quantity of inorganic carbonates, necessitating the measurement of CO_2 levels to accurately assess the mineral composition of these coals with elevated carbonate concentrations. Understanding the coal's carbonate composition is imperative to rectify the carbon measurement. Expressing the carbonate amount as precise amounts of calcium carbonates, magnesium ions carbonate, and other components would require complex analyses. Therefore, it is conventional and satisfactory for analytical objectives to denote it solely in atmospheric CO_2 terms, as more elaborate methods are unnecessary.

Coal contains arsenic and selenium in a few parts per million concentrations. During coal combustion, a variable amount of these elements is either released or maintained in the ash, largely dependent on the combustion circumstances and the characteristics of the ash from the coal.

2.3. SUMMARY

Coal is a type of organic sedimentary rock that is primarily composed of carbonized plant matter. It forms the accumulation and preservation of plant remains in swampy environments over millions of years. Coal is considered a fossil fuel and is one of the world's most abundant and widely used energy sources. There are four main type of coal, which differs in their carbon content, energy content and geological origin including peat, lignite, bituminous coal, and anthracite. Efforts have been made to develop cleaner and more sustainable alternatives to coal as an energy source, such as renewable energy technologies like solar, wind, and hydropower.

REFERENCES

1. Anderson, K. B., & Mackay, G., (1990). *International Journal of Coal Geology, 16*, 327.
2. ASTM., (2011). Test method for forms of sulfur in coal (ASTM D2492). *Annual Book of ASTM Standards, Section 05.05*. American Society for Testing and Materials, West Conshohocken, PA.
3. Bailey, A. M., Sherrill, J. F., Blackson, J. H., & Kosters, E. C., (1990). Sulfur and pyrite in precursors for coal and associated rocks: A reconnaissance study of three modern sites. *Geochemistry of Sulfur in Fossil Fuels; ACS Symposium Series* (Vol. 429, pp. 186–203). American Chemical Society, Washington, DC, Chapter 10.
4. Baker, E. W., (1969). In: Eglinton, G., & Murphy, M. T. J., (eds.), *Organic Geochemistry: Methods and Results*. Springer-Verlag, New York, Chapter 19.
5. Bend, S. L., (1992). *Fuel, 71*, 851.
6. Bend, S. L., Edwards, I. A. S., & Marsh, H., (1991). *Fuel, 70*, 1147.
7. Berkowitz, N., (1988). In: Ebert, L. B., (ed.), *Polynuclear Aromatic Compounds*. Advances in Chemistry Series No. 217. American Chemical Society, Washington, DC, Chapter 14.
8. Bodzek, D., & Marzek, A., (1981). *Fuel, 60*, 47.
9. Bonnett, R., Czechowski, F., & Hughes, P. S., (1991). *Chemical Geology, 91*, 193.
10. Bouska, V., (1981). *The Geochemistry of Coal*. Elsevier, Amsterdam, the Netherlands.
11. Branthaver, J. F., (1990). In: Speight, J. G., (ed.), *Fuel Science and Technology Handbook*. Marcel Dekker, Inc., New York, Chapter 3.
12. Bredenberg, J. B., Huuska, M., & Vuori, A., (1987). In: Volborth, A., (ed.), *Coal Science and Chemistry*. Elsevier, Amsterdam, the Netherlands.
13. Breger, I. A., (1958). *Economic Geology, 53*, 823.
14. Breger, I. A., (1977). In: *Future Supply of Nature-Made Petroleum and Gas* (p. 913). Pergamon, New York.
15. Casagrande, D. J., & Seifert, K., (1977). *Science, 195*, 675.
16. Casagrande, D. J., (1987). Sulfur in peat and coal. In: Scott, A. C., (ed.), *Coal and Coal-Bearing Strata: Recent Advances* (pp. 87–105). Geological Society Special Publication No. 32. Presentation to Geological Society of London, London, U.K.
17. Casagrande, D. J., Idowu, G., Friedman, A., Rickert, P., Siefert, K., & Schlenz, D., (1979). *Nature, 282*, 599.

18. Chou, C. L., (1990). In: *Geochemistry of Sulfur in Fossil Fuels*. Symposium Series No. 429. American Chemical Society, Washington, DC, Chapter 2.
19. Christman, R. F., & Oglesby, R. T., (1971). In: Sarkanen, K. V., & Ludwig, C. H., (eds.), *Lignins: Occurrence, Formation, Structure, and Reactions*. Interscience/John Wiley & Sons, Inc., New York, Chapter 18.
20. Cooper, B. S., & Murchison, D. G., (1969). In: Eglinton, G., & Murphy, M. T. J. (eds.), *Organic Geochemistry: Methods and Results*. Springer-Verlag, New York, Chapter 29.
21. Dalton, D. R., (1979). *The Alkaloids: The Fundamental Chemistry, A: Biogenic Approach*. Marcel Dekker, Inc., New York.
22. Davidson, R., (1994). *Nitrogen in Coal*. Report No. IEAPER/08. International Energy Agency, London, U.K.
23. Deno, N. C., Greigger, B. A., & Stroud, S. G., (1978). *Fuel, 57*, 455.
24. Diessel, C. F. K., (1980). *Coal Geology*. Australian Mineral Foundation, Adelaide, South Australia, Australia.
25. Duty, R. C., Hyatsu, R., Scott, R. G., Moore, L. P., Winans, R. E., & Studier, M. H., (1980). *Fuel, 59*, 97.
26. Ekwenchi, M. M., Imobighe, G. A., & Adejumo, M. A., (1991). *Fuel Processing Technology, 28*, 239.
27. Elliott, M. A., (1981). *Chemistry of Coal Utilization, Second Supplementary Volume*. John Wiley & Sons, Inc., New York.
28. Francis, W., (1961). *Coal: Its Formation and Composition*. Edward Arnold Ltd., London, U.K.
29. Given, P. H., (1975). In: Eglinton, G. (ed.), *Environmental Chemistry* (Vol. 1). The Royal Society of Chemistry, London, England, Chapter 3.
30. Greer, R. T., (1979). *Scanning Electron Microscopy, 1*, 377.
31. Hare, P. E., (1969). In: Eglinton, G., & Murphy, M. T. J., (eds.), *Organic Geochemistry: Methods and Results*. Springer-Verlag, New York, Chapter 18.
32. Hatcher, P. G., & Lerch, H. E. III., (1991). In: Schobert, H. H., Bartle, K. D., & Lynch, L. J. D., (eds.), *Coal Science II, Symposium Series No. 461*. American Chemical Society, Washington, DC, Chapter 2.
33. Hendricks, T. A., (1945). In: Lowry, H. H., (ed.), *Chemistry of Coal Utilization* (Vol. 1). John Wiley & Sons, New York, Chapter 1.
34. Hessley, R. K., (1990). In: Speight, J. G., (ed.), *Fuel Science and Technology Handbook*. Marcel Dekker, Inc., New York.
35. Hessley, R. K., Reasoner, J. W., & Riley, J. T., (1986). *Coal Science*. John

Wiley & Sons, Inc., New York.

36. Hindmarsh, C. J., Wang, W., Thomas, K. M., & Crelling, J. C., (1994). The release of nitrogen during the combustion of macerals, microlithotypes, and their chars. *Fuel, 73*, 1229–1234.
37. Huck, G., & Karweil, J., (1955). *Brennstoff Chemie, 36*, 1.
38. Hyatsu, R., Winans, R. E., & McBeth, R. L., (1979). Lignin-like polymers in coal. *Nature, 278*(5699), 31.
39. Hyatsu, R., Winans, R. E., McBeth, R. L., Scott, R. G., Moore, L. P., & Studier, M. N., (1981). In: Gorbaty, M. L., & Ouchi, K., (eds.), *Coal Structure*. Advances in Chemistry Series No. 192. American Chemical Society, Washington, DC, Chapter 9.
40. ISO., (2011). *Hard Coal – Determination of Forms of Sulfur (ISO 157)* (p. 331). International Standards Organization, Geneva, Switzerland.
41. Karweil, J., (1956). Z Deutsche. *Geologische Gesellschaft, 107*, 132.
42. Keleher, (1995). *The Significance of Nitrogen in the Coal Combustion Process*. Department of Mines and Energy, Queensland Department of Minerals and Energy, Brisbane, Queensland, Australia.
43. Kreulen, D. J. W., (1948). In: Nijgh & van Ditmar, Rotterdam, N. V., (eds.), *Elements of Coal Chemistry*. The Netherlands.
44. Lee, M. L., Novotny, M. V., & Bartle, K. D., (1981). *Analytical Chemistry of Polycyclic Aromatic Compounds*. Academic Press, Inc., New York, Chapter 2.
45. Lowry, H. H., (1945). *Chemistry of Coal Utilization*. John Wiley & Sons, Inc., New York.
46. Lowry, H. H., (1963). *Chemistry of Coal Utilization* (Supplementary Volume). John Wiley & Sons, Inc., New York.
47. Lu, D., Anthony, E. J., & Tan, Y., (2007). Mercury removal from coal combustion by Fenton reactions. *Fuel, 86*, 2789–2797.
48. Lukoshko, E. S., Bambalov, N. H., & Smychnik, T. P., (1979). The change in composition of lignin in the process of peat formation. *Solid Fuel Chemistry (USSR), 13*(3), 132.
49. Luo, B. J., Wang, Y. X., Meng, Q. X., Yang, X. H., Li, X. Y., & Cheng, N., (1991). *Science in China (Series B), 34*, 363.
50. Masuku, C. P., (1992). *Journal of Analytical and Applied Pyrolysis, 23*, 195.
51. Meinschein, W. G., (1969). In: Eglinton, G., & Murphy, M. T. J., (eds.), *Organic Geochemistry: Methods and Results*. Springer-Verlag, New York, Chapter 13.

52. Middleton, S. P., Patrick, J. W., & Walker, A., (1997). The release of coal nitrogen and sulfur on pyrolysis and partial gasification in a fluidized bed. *Fuel, 76*, 1195–1200.
53. Moore, E. S., (1940). *Coal: Its Properties, Analysis, Classification, Geology, Extraction, Uses, and Distribution*. John Wiley & Sons, Inc., New York.
54. Mullins, O. C., Mitra-Kirtley, S., Van Elp, J., & Cramer, S. P., (1993). Molecular structure of nitrogen in coal from XANES spectroscopy. *Applied Spectroscopy, 47*, 1268–1275.
55. Murchison, D., & Westoll, T. S., (1968). *Coal and Coal Bearing Strata*. Elsevier, New York.
56. Nishioka, M., & Lee, M. L., (1988). In: Ebert, L. B. (ed.), *Polynuclear Aromatic Compounds*. Advances in Chemistry Series D No. 217. American Chemical Society, Washington, DC, Chapter 14.
57. Orheim, A., (1979). In: Argall, Jr. G. O. (ed.), *Coal Exploration: Proceedings of the Second International Coal Exploration Symposium*. Miller Freeman Publications, Inc., San Francisco, CA, Chapter 14.
58. Palmer, S. R., Hippo, E. J., Crelling, J. C., & Kruge, M. A., (1992). Characterization and selective removal of organic sulfur from Illinois Basin coals. *Coal Preparation, 10*, 93–106.
59. Palmer, S. R., Kruge, M. A., Hippo, E. J., & Crelling, J. C., (1994). Speciation of heteroatoms in coal by sulfur- and nitrogen-selective techniques. *Fuel, 73*, 1167–1172.
60. Parker, P. L., (1969). In: Eglinton, G., & Murphy, M. T. J., (eds.), *Organic Geochemistry: Methods and Results*. Springer-Verlag, New York, Chapter 14.
61. Richardson, J. S., & Holba, A., (1987). In: Volborth, A. (ed.), *Coal Science and Chemistry* (p. 343). Elsevier, Amsterdam, the Netherlands.
62. Roberts, J. D., & Caserio, M. C., (1964). *Basic Principles of Organic Chemistry*. Benjamin Book Publishers, Inc., New York.
63. Sarkanen, K. V., & Ludwig, C. H., (1971). *Lignins: Occurrence, Formation, Structure, and Reactions*. Interscience/John Wiley & Sons, Inc., New York.
64. Schiller, J. R., (1977). *Analytical Chemistry, 49*, 2292.
65. Schmidt, R. A., (1979). *Coal in America: An Encyclopedia of Reserves, Production, and Use*. McGraw-Hill, New York.
66. Schweighardt, F. K., White, C. M., Friedman, S., & Schultz, J. L., (1977). *Preprints. Division of Fuel Chemistry* (Vol. 22, No. 5, p. 124). American

Chemical Society.

67. Schwendinger, R. B., (1969). In: Eglinton, G., & Murphy, M. T. J., (eds.), *Organic Geochemistry: Methods and Results*. Springer-Verlag, New York, Chapter 17.
68. Scouten, C. S., (1990). In: Speight, J. G., (ed.), *Fuel Science and Technology Handbook*. Marcel Dekker, New York, Chapter 28.
69. Smith, G. C., & Cook, A. C., (1980). Coalification paths of exinite, vitrinite, and inertinite. *Fuel, 59*, 641.
70. Snowdon, L. R., (1991). *Organic Geochemistry, 17*, 743.
71. Speight, J. G., (1980). *The Chemistry and Technology of Petroleum* (1st edn.). Marcel Dekker Inc., New York.
72. Speight, J. G., (2005). *Handbook of Coal Analysis*. John Wiley & Sons, Inc., Hoboken, NJ.
73. Speight, J. G., (2007). *The Chemistry and Technology of Petroleum* (4th edn.) Taylor & Francis Group, Boca Raton, FL.
74. Stach, E., Taylor, G. H., Mackowsky, M.-Th., Chandra, D., Teichmuller, M., & Teichmuller, R., (1982). *Textbook of Coal Petrology* (3rd edn.). Gebruder Bornfraeger, Stuttgart, Germany.
75. Streibl, M., & Herout, V., (1969). In: Eglinton, G., & Murphy, M. T. J., (eds.), *Organic Geochemistry: Methods and Results*. Springer-Verlag, New York, Chapter 16.
76. Stutzer, O., (1940). *Geology of Coal*. University of Chicago Press, Chicago, IL.
77. Sunavala, P. D., (1990). *URJA, 28*(2), 31.
78. Swain, F. M., (1969). In: Eglinton, G., & Murphy, M. T. J., (eds.), *Organic Geochemistry: Methods and Results*. Springer-Verlag, New York, Chapter 15.
79. Teichmuller, M., & Teichmuller, R., (1979). The diagenesis of coal (coalification). In: Larsen, G., & Chilingarian, G. V., (eds.), *Diagenesis in Sediments and Sedimentary Rocks*. Elsevier, New York, Chapter 5.
80. Terres, E., (1931). Contributions to the origin of coal and petroleum. *Proceedings – Third International Conference on Bituminous Coal, 2*, 797.
81. Tewalt, S. J., Bragg, L. J., & Finkelman, R., (2001). *Mercury in US Coal: Abundance, Distribution, and Modes of Occurrence, Fact Sheet FS-095-01*. United States Geological Survey, Washington, DC.
82. Van Krevelen, D. W., (1961). *Coal: Typology, Chemistry, Physics, Constitution*. Elsevier, Amsterdam, the Netherlands.

83. Wang, W., Brown, S. D., Thomas, K. M., & Crelling, J. C., (1994). Reactivity and nitrogen release from a rank series of coals during temperature programmed combustion. *Fuel, 73*, 341–347.
84. Wawrzynkiewicz, W., & Sablik, J., (2002). Organic sulfur in the hard coal of the stratigraphic layers of the Upper Silesian coal basin. *Fuel, 81*, 1889–1895.
85. Weiss, U., & Edwards, J. M., (1980). *The Biosynthesis of Aromatic Compounds*. John Wiley & Sons, Inc., New York.
86. White, C. M., (1983). In: Bjorseth, A., (ed.), *Handbook of Polycyclic Aromatic Hydrocarbons*. Marcel Dekker, Inc., New York, Chapter 13.
87. Wollrab, V., & Streibl, M., (1969). In: Eglinton, G., & Murphy, M. T. J., (eds.), *Organic Geochemistry: Methods and Results*. Springer-Verlag, New York, Chapter 24.

CHAPTER

3 Preparation, Transportation, and Storage of Coal

Contents

INTRODUCTION

Coal preparation, also known as coal cleaning, involves eliminating impurities from fossil fuels, including sulfuric acid, ashes, and rock, to enhance its economic worth. The classification of coal preparation methods can be broadly divided into two categories: physical cleaning and chemical cleaning. Currently, the predominant methods employed for separating fossil fuels from impurities are clean-up processes that rely on variations in density. These processes involve mechanical means. At present, there are ongoing efforts to create chemical coal cleaning techniques. However, their efficacy and economic feasibility remain uncertain.

The primary aims of coal cleaning procedures are to decrease, inside predetermined thresholds, the dimensions, moisture content, ash content, and sulfur content. The properties of coal significantly impact the feasibility of coal cleaning and the appropriate methods to be employed. The coal's rank, a multifaceted characteristic that describes the nature and characteristics of coal, is a significant factor in assessing the viability and scope of coal cleaning.

The impact of coal quality on coal combustion, particularly on various aspects of the plant's operation, has become widely acknowledged. The variables of rank, minerals content (ash content), sulfuric acid content, and water content are considered crucial factors for assessing a substance's combustibility concerning its heating value and reactivity. Furthermore, it is worth noting that the size of coal particles can significantly impact their combustion behavior during power plant activities, even though it may not always be acknowledged as a type of cleaning or beneficiation. Therefore, it is imperative to undertake any kind of pretreatment, such as cleaning, before utilizing coal.

The primary objective of coal processing is to enhance the caliber of coal for its immediate application in producing steam and electricity or distinctive purposes, including chemical fuel, feed to liquefaction, and gasification. The presence and amount of impurities in fossil fuels are widely recognized as the primary factors that impose restrictions on coal utilization. There exists a significant variability among coals. The selection of coal processing technology and the degree of beneficiation primarily hinges on the coal variety, mining methods, and how to use clean coal.

Typically, the yield of a coal mine comprises unprocessed coal (known as run-of-mine or ROM, or read coal), characterized by higher moisture content and finer particle size. Moreover, contemporary mining practices, which involve the excavation of inferior quality coal reserves, employment of advanced and non-discriminatory mining technologies, and a great deal of water for dust suppression, have led to an increase in the proportion of contaminants in the ROM coal. The declining quality of ROM coals has led to a heightened need for effective coal processing technology, as there is now a greater need for high-quality coal that conforms with market and environmental requirements.

The mineral has significant content and exhibits significant variability, ranging from 3% to 60% in mineral ash across various mining sites. The majority of ash is typically introduced to the roofing or bottom inside my own or from partings – which are small edges of slate – within the coal seam. The unrelated ash, which refers to the mineral content in coal that is not inherent to it, has a specific gravity greater than 1.80. The mineral matter after removing the extraneous ash is intrinsic to the coal. The presence of mineral-laden ash in coal is directly proportional to its density.

Coal contains sulfur compounds in sulfates, natural sulfur, and pyrites, iron sulfides. Sulfates are typically found in minimal concentrations and are generally deemed inconsequential. The molecular binding of organic sulfur within coal precludes its removal through conventional coal preparation techniques. Pyrite is commonly found in nodules or may be closely intermixed with coal. The efficacy of sulfur reduction in coal is contingent upon multiple factors, including the concentration of pyrites in fossil fuels, the degree of integration between pyrites and coal, and the level of coal preparation. It is worth noting that coal-getting-ready plants can only eliminate a fraction of the total pyritic sulfur present in coal. The dimensions of coal fragments exhibit a wide spectrum, extending to the magnitude of extraneous substances, such as a boulder dislodged from the mine ceiling or a metallic fastener. Occasionally, sizable coal chunks sourced from an exceptionally rigid stratum are incorporated. Over the coal extraction process, various foreign substances are introduced, including but not limited to roof screws, ties, car tires, timber, struck wires, along with cutting bits.

Did You Know?

In ancient times, coal was used for medicinal purposes, including treating ailments like skin diseases and digestives issues.

Coal getting ready serves multiple objectives. A significant objective is to enhance the calorific value of coal by physically eliminating contaminants.

Frequently, it is necessary to identify an audience for the product. Refuse materials can constitute up to 60% of the ROM coal extracted from contemporary mining operations.

Most traditional coal cleaning procedures employ gravitational techniques for larger particle sizes and surface treatment techniques for smaller particle sizes. The choice of equipment, particularly for the smaller sizes, is contingent upon the mining technique employed, the degree of coal hardness, how to distribute, and the number of sizes present.

3.1. COAL PREPARATION

The process of coal getting ready serves multiple objectives. A significant objective is to enhance the calorific value of coal by physically eliminating contaminants. Frequently, it is necessary to identify the market for the product. Refuse materials can constitute up to 60% of the ROM coal extracted from a contemporary mining operation.

The correlation between the energy density of coal and its ranking (degree of coalification) is established (as shown in Table 3.1). It is subject to the influence of nonfuel components such as minerals and moisture. Hence, the principal aim of coal washing is to optimize the recuperation of the calorific value of coal while conforming to established criteria for ash, moisture, and sulfur levels.

Did You Know?

The process of converting coal into coke is known as coking, and it involves heating coal in the absence of oxygen.

In addition, it is noteworthy that transportation expenses are commonly assessed based on a ton-mile metric, which needs to differentiate between the coal's inherent substance and its moisture content. Consequently, it is advantageous to eliminate any superfluous mineral makeup and water before shipping, as this can lower transportation expenses for a "substandard" coal grade while simultaneously providing a more energy-dense material to the end user.

In essence, the economic worth of a given coal type is determined by its grade concerning a particular intended application. The grade of coal pertains to the quantity of mineral matter contained within the coal and indicates its overall quality. The two parameters of interest are the sulfur content and the ash integration temperature, a metric used to assess the ash's behavior under high-temperature conditions.

Table 3.1. Typical Properties of Different Rank Coals

Cole Type	Carbon (%)	Hydrogen (%)	Limits of Volatile Matter (%)	Fixed Carbon (%)	Calorific Value (Btu/lb)
Lignites	73–78	5.2–5.6	45–50	50–55	<8,300
Subbituminous	78–82.5	5.2–5.6	40–45	55–60	8,300–11,500
High-volatile bituminous	82.5–87	5.0–5.6	30–40	60–70	11,500–14,000
Medium-volatile bituminous	87–92	4.6–5.2	20–30	70–80	>14,000
Low-volatile bituminous	91–92	4.2–4.6	15–20	80–85	>14,000
Anthracite	95–98	2.9–3.8	5–10	91–95	>14,000

Source: John F. Fryer, Creative Commons Lic.

The presence and amount of minor elements also determine the grading of coal. While there is a lack of legal binding based on fossil fuel grade, coal quality remains a crucial factor for coal consumers.

The transportation of lignites over long distances, containing over 33% water, can significantly increase the starting mine-mouth costs calculated based on energy consumption. Possible academic rewrite: Transportation costs could be affected by a trade-off when dealing with low-rank coal, especially if its sulfur content is low enough to reduce the expenses of stack gas disposal.

Reducing coal cleaning requirements can be reduced by selecting appropriate mining techniques. While some mining operations incorporate methods to decrease the size of large coal, washing ROM, coal is typically a distinct process conducted as a surface functioning at the mine's entrance. The term "coal preparation" encompasses sizing techniques such as crushing and breaking and everything handling and processing methods necessary to ready the coal for commercial distribution.

The physical coal cleaning procedures employed in coal cleaning facilities exhibit variability, albeit broadly categorized into four fundamental stages: initial planning, fine fossil fuels processing, rough coal processing, and final preparation.

During the preliminary stage of coal sanitation, the unprocessed coal is discharged, stockpiled, transported, comminuted, and segregated through screening into two distinct coal fractions: coarse and fine. Subsequently, the size parts are transported to their corresponding purification procedures.

Separating pollutants in fine coal analysis and coarse in texture coal processing involves comparable operations and equipment. The main distinction lies in the degree of severity exhibited by the operational parameters. Most coal cleaning methods employ fluidization techniques, such as upward electrical currents or impulses of water, that circulate a bed of smashed coal and impurities. The upward movement of the coal particles with lower density results in their separation from the bed and subsequent extraction from the upper region. The process involves the separation of denser impurities from the lower portion of the mixture. Coal that has undergone wet cleaning must subsequently dry during the final preparation stages.

Utilizing the last preparation procedures, the moisture content in coal consequently mitigates the issues of weight and freezing while simultaneously augmenting its heating potential. The initial processing stage involves dewatering, wherein a significant proportion of the aqueous component is separated through screens, thickening agents, and cyclones. Typically, the second phase involves thermal drying, which can be accomplished using one of the following dryers: fluidized bed, lightning, and multi-louvered. Drying coal in a fluid-bed dryer involves suspending the coal particles above a perforated plate, subjected to rising hot gases. The flash dryer operates by introducing coal into a flow of heated gases, resulting in rapid and immediate desiccation. The desiccated coal and moist gases are conveyed upwards through drying columns and subsequently segregated via a cyclonic mechanism. The multi-louvered drier heated gases through a descending stream of coal, which is subsequently elevated by a conveyor system featuring specialized flights.

The elimination of free sulfur present in coal can solely be accomplished through chemical processing or combustion, neither of which constitutes a coal-preparing procedure. The partial removal of pyrites during washing procedures is attributed to their weight, which enables them to be eliminated alongside the ash. The efficacy of the procedures in eliminating pyrites is limited to a range of 30% to 60%. This is because certain pyrites remain attached to the raw material and are not separated during the process. Additionally, the pyrites found in an individual piece may not be significant enough to warrant rejection.

The fundamental objectives of coal preparation for commercial purposes are contingent upon the characteristics of the initial coal product (as outlined in Table 3.2). In essence, these objectives involve the reduction of coal particle size and regulating particle size within the parameters dictated by logistics and handling requirements.

Table 3.2. General Methods of Coal Preparation and Levels of Cleaning

Level	Raw Coal Weight (%)	Raw Coal Btu Content (%)	Reduction Potential		Comments
			Ash	Sulfur	
1	98–100	99–100	None to minor	None	Crushing and breaking of raw coal to 3 in. size.
2	75–85	90–95	Fair to Good	None to minor	Coarse coal cleaning of 3× 3/8 in. coal.
3	60–80	80–90	Good	Fair	Moderate coal cleaning of 3 in. × 28-mesh coal.
4	60–80	80–90	Good to excellent	Fair to good	Fine coal cleaning of 3 in. × 0-mesh coal.

5	60–80	85–95	Deep-cleaned coal: excellent; middle-cleaned coal: none to fair.	Multiple-stream coal preparation of two cleaned coal products: "deep-cleaned" coal and "middle grade" coal.	–

Source: James G. Speight, Creative Commons License.

The process involves optimizing the use of the material and eliminating superfluous mineral components to achieve a level of satisfaction for the consumer while adhering to the prescribed standards. The procedure above is commonly denoted as the regulation of ash concentration.

3.2. SIZE REDUCTION

Pretreatment involves the fragmentation, comminution, and classification of the ROM coal to establish a consistent raw coal feed with a specified maximum particle size. This reduces the generation of excessively fine material resulting from over-crushing or handling. The terms before order are frequently utilized interchangeably.

Most traditional coal cleanup facilities employ gravity-based techniques for the larger size fractions and surface-based methods for the smaller particle sizes. The choice of equipment, particularly for the smaller particle sizes, is contingent upon the mining technique employed, the level of coal toughness, and the quantity and distribution of particle sizes.

The initial procedures executed on ROM coal involve eliminating extraneous ferrous materials and the reduction of dimensions to enable mechanical manipulation. The initial step in the purification process of ROM coal involves subjecting it to an intense magnet, typically suspended above an arriving strap conveyor, to extract iron contaminants from the coal.

The magnetic component pursues the circuit breaker, albeit consistently preceding the screen-crushing apparatus. Subsequently, the coal is directed towards the breaker, a capacious cylindrical structure equipped with internal lifting blades. The shell surrounding the breaker is punctured with apertures measuring 2 to 8 inches in diameter, facilitating diminutive substances' transit.

The breaker undergoes horizontal axis rotation and functions by receiving substance at one end, allowing it to tumble through the shell's perforations and facilitating the exit of hard, large, unaltered material from the device's rear end. The four-inch diminutive material is directed towards the cleaning facility, while the bulky and discarded material is deposited into a receptacle for subsequent removal.

Most commercial networks employ dense-media vessels or jigs for the largest size, typically +3/8 in., packed the media hurricanes, focusing tables, or tools for the size range of 3/8 in. × 28-mesh, water-only cyclones, or spiraling shapes, occasionally flotation for the size range of 28 × 100-mesh, and flotation for the smallest size range of 100 mesh.

The utilization of heavy-medium hurricanes is widespread in the purification of coals of moderate size, ranging from 0.2 to 10 mm. Coarse products are subjected to dewatering through testing and centrifugal drying centrifugal pumps, and occasionally, thermal dryers are employed to decrease the moisture level of the finer sizes. Most coal cleaning plants specializing in metallurgical coal employ thermal dryers. The extracted coal exhibits a greater degree of softness and friability, resulting in a more refined size distribution facilitated by the mining machinery. To satisfy the demands of end users, coals intended for metallurgical purposes necessitate meticulous processing and drying.

Furthermore, flotation is a ubiquitous technique employed in these circuits owing to the substantial amount of coal and requisite end product quality, characterized by low ashes and low sulfur content. In contrast, certain steam coals, particularly those with greater hardness (as indicated by a low hard grooved index) and certain coals derived from surface mining operations, exhibit reduced levels of the 100-mesh dimensions. In numerous plant species, the amount of 100-mesh size is so minimal that it is deemed uneconomical to retrieve and is consequently dispatched for disposal.

Reducing the dimensions of coal is a crucial factor in maximizing the utilization of ROM coal for various purposes, including electricity generation, coke manufacturing, and the creation of artificial fuels. ROM coal refers to the coal obtained directly from the mine, regardless of the specific mining technique employed, such as removal, auger mining, ongoing extraction, short- or long-barrier mineral extraction, or any other technique currently utilized.

Due to variations in mining procedures and the commencement of size reduction at the face of the coal throughout mining, the coal extracted will display distinct properties. The coal extraction method significantly impacts the extracted coal's size and size uniformity. Consequently, it is typically necessary to perform a certain level of size reduction or size regulation before the ultimate consumption of coal. The quantity of stages in reducing coal size depends on the coal's particular application and state.

Reducing the volume of coal is a crucial factor in maximizing the utilization of ROM coal for various purposes, such as power generation, coke manufacturing, and artificial fuel production. The term "ROM coal" refers to the coal obtained from the mining procedure without undergoing treatment. This includes various mining techniques such as removal, augering, ongoing mining, instantaneous or long-wall mineral extraction, or any other currently used technique. After undergoing screening or crushing procedures, the ROM coal is designated as raw coal.

The decrease in the ratio pertains to the proportion between the highest size of the input material and the highest size of the output material or the proportion among the sizes of the input and output materials at which a designated proportion of the substance traverses. In the context of a reduction in size units, it is typical for those that yield a product top size exceeding 1 mm to exhibit size reduction ratios of approximately 8–1. Conversely, in units where the final item's highest size is less than 1 mm, size decrease ratios may span from 200 to 1 and beyond.

Typically, the size of ROM coal is decreased by utilizing four distinct types of machinery: rotary breaks, roll crushing equipment, hammer mills (ring mills), and impactors. The optimal outcome of the crushing process is attaining the intended product size within a solitary stage. Nevertheless, there exist instances wherein the diminution of coal dimensions necessitates the utilization of several stages of size reduction apparatus. The ultimate step in the coal before-treatment procedure involves the segregation of coal into size fractions suitable for efficient handling by the diverse process waterways.

It is relevant to explain the diverse terminologies associated with coal pretreatment. The maximum size threshold for the crushed material is determined by the minimum aperture size of the sieve that can retain a maximum of 5% of the total samples. The expression "nominal top size" pertains to the output of the size reduction process and denotes the minimum aperture of a sieve that can retain between 5% and 20% of the specimen. The reduction ratio pertains to the proportion between the uppermost size of the input material and the uppermost size of the output material or the proportion among the sizes of the input and output materials at which a specific percentage of the substance traverses. The comminution (reduction in top size) of coal through mechanical action significantly impacts the architectural layout of size reduction units or operations. The ease with which coal can be ground is commonly referred to by the grindability indices of the coal. The Hardgrove Grindability Index (HGI) is the predominant measure employed with coal particle size reduction. The HGI are established by grinding 50 g of 16 × 30-mesh desiccated coal in a cylindrical ball and ethnicity mill that adheres to standardized procedures. Subsequently, the sample is extracted and subjected to sieving to ascertain the quantity of material (W) that can traverse a 200-mesh sieve. The HGI is inversely proportional to the fuel's durability during crushing, implying that an increase in resistance leads to a reduction in the HGI. The concept of specific energy pertains to the amount of energy needed per unit of material processed to achieve the desired particle size of a given coal feed.

The relationship between the specific energy required for reducing the size and the grindability indices of coal is directly correlated and is further influenced by the lower ratio. The energy demand for coal size reduction positively correlates with the reduction ratio and throughput. The energy demand for a typical coal throughput fluctuates based on the decrease in the ratio, as illustrated in Figures 3.1 and 3.2. Moreover, the correlation between the particular form of energy and the grindability is contingent on the type of equipment employed, as demonstrated in Figure 3.3.

The presence of humidity within coal is a multifaceted concern that extends beyond the mere dryness of the coal and the economic feasibility of drying it for transportation purposes. The consideration of moisture content is imperative when assessing the energy demands of a smaller apparatus. An excessive amount of moisture, especially on the surface, may affect fines removal effectiveness due to the accumulation of fines. Additionally, the creation of emulsions can pose a challenge when using the selective accumulation process.

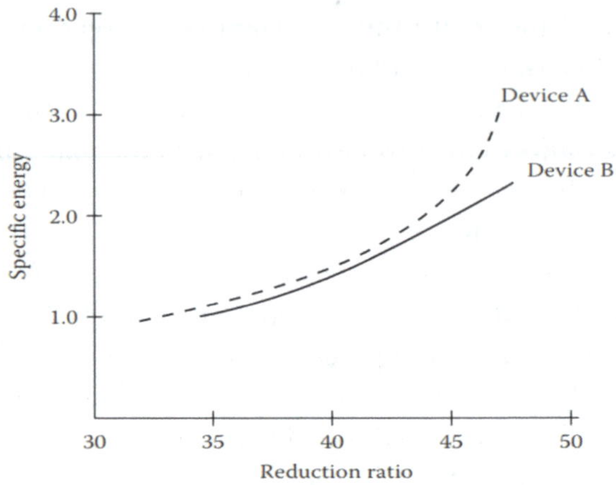

Figure 3.1. *Relationship of specific energy to size reduction for two coal cleaning devices.*

Source: Meyers, Creative Commons License.

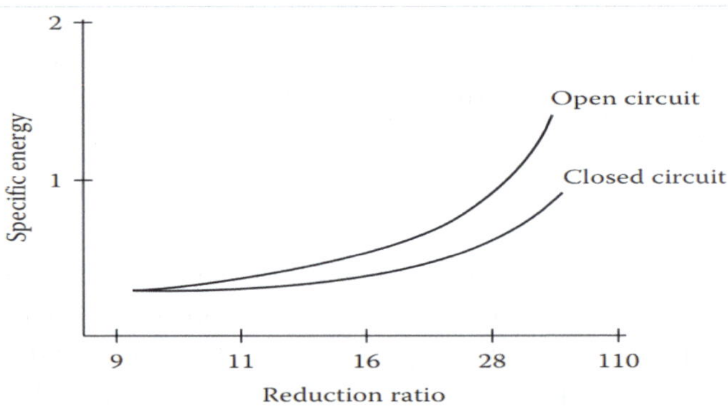

Figure 3.2. *Relationship of specific energy to size reduction for open and closed circuits.*

Source: Marcel Dekkers, Creative Commons License.

Did You Know?

The world's first steam-powered locomotive, developed by George Stephenson in 1814, was designed to transport coal from mines to a canal for further transportation.

3.3. COAL CLEANING

The ash in a kilogram of ROM coal typically ranges from 5% to 40%. In comparison, the amount of sulfur is approximately 0.34%, which is subject to variation based on the geological factors and mining methods employed. The effective and ecologically sound use of coal often necessitates the removal of excessive impurities through coal cleaning. Coal cleanup is a prevalent practice at mines located in the Eastern and Midwestern regions of the US.

The predominant techniques employed in contemporary coal cleaning processes are based upon separation by force, as biological and chemical techniques are often deemed cost-prohibitive. The density separation process is commonly employed to purify larger-sized coal particles. In contrast, techniques that rely on surface properties are typically favored for the purification of smaller-sized coal particles. The density-based techniques involve the introduction of coal particles into a liquid medium, followed by applying either gravitational or centrifugal forces to effect the separation of the organic-rich (float) stage into the rich-in minerals (sink) phase. Typically, contemporary techniques for coal purification in the commercial sector rely on mechanical separation, as chemical-based and biological alternatives are often deemed cost-prohibitive. In general, density separation is commonly employed to purify large-sized coal particles. In contrast, techniques based on surface properties are typically favored for cleaning smaller coal particles. The density-based techniques involve the introduction of coal particles into a liquid medium, followed by applying either gravitational or centrifugal force to effect the separation of the organic-rich (float) stage into the rich-in minerals (sink) phase. Separating materials based on density is commonly achieved in commercial settings through various equipment such as tools, minerals spiraling shapes, concentrating tables, water-based hurricanes, and heavy-media dividers. The efficacy of density-based washing circuits is evaluated through laboratory float-sink examinations. These surface property-based procedures involve admiring ground coal with water and adding a minute quantity of collector reagent to enhance the hydrophilic nature of coal surfaces. Following this, the process involves the introduction of air bubbles within the presence of the frother agent, which facilitates the transportation of coal fragments to the upper layer of the slurry, thereby segregating them away from hydrophilic particles of minerals. Achieving exterior property-based washing in commercial settings involves using foam that sits or column floating techniques.

3.3.1. Effect of Composition and Rank

Coal in its pure form primarily comprises carbon, accompanied by relatively smaller proportions of the element hydrogen, oxygen, nitrogen, and sulfur, which is present inside the organic matrix. Additionally, mineral contaminants are also present in coal. The organic matrix transforms after coal combustion into gaseous burning goods; mineral impurities persist as a residual substance called "ash."

Notably, the ash from combustion does not comprise the mineral content initially inherent within the coal but rather the remnants that persist after the elements have converted into diverse elemental oxides.

Alterations arising from the breakdown of mineral constituents in this way do not furnish any input to the thermal energy content of the coal. Instead, they typically diminish it. Adjustments to the measured values are necessary when precise calorie content must be determined. The magnitude of these adjustments is relatively small, and it is commonly deemed satisfactory (albeit somewhat imprecise) to employ the ash content as an indicator of the mineral contaminants existing in coal, given its ease and expeditiousness of determination.

The pyrolyzing coals involve the subsequent removal of hydrogen and oxygen, resulting in near-complete removal. Nevertheless, the carbon residue retains minor amounts of nitrogen along with sulfur. The coal rank classification system is based on typical amounts of carbon, hydrogen, volatile matter, and residual carbon for different classes of coal, as presented in Table 3.1. These parameters are used to describe the rank of coal.

The chemical composition of coal is intricate; however, compelling indications suggest that its nitrogen is wholly encompassed within the organic matrix, with minimal or no contradictory evidence. Conversely, oxygen and sulfur are present in compounds inorganic and organic.

Oxygen is frequently deemed insignificant in coal purification due to its non-polluting nature. It is important to take into account the impact of oxygen upon the destiny of sulfur and nitrogen compounds, as well as its influence on the formation of corrosive water-based acids when water, a by-product of hydrocarbons burning that contains oxygen, compresses with additional products upon the cooler areas of combustion systems of all kinds.

Sulfur is a distinctive instance as it is regarded and substantively recognized as a more significant contaminant than oxygen. Sulfur manifests in diverse configurations and is ubiquitously dispersed within the matrix of organic matter in mineral content. Organic sulfur is present within the organic composition of coal, while pyritic sulfuric acid is present in the form of distinct pyrite (Fe) particles. Furthermore, sulfates can be detected in certain minerals. To summarize, most of the sulfur, whether in organic or inorganic form, found in coal can manifest as gaseous byproducts during combustion.

3.3.2. Methods

The techniques utilized for eliminating impurities from coal are contingent upon the physical dissimilarities between the contaminants and the coal. A notable instance of this is the variation in specific gravity.

Typically, coal-getting-ready facilities employ gravity-based processing machinery to segregate the undesired materials from the coal. Most extraneous contaminants extracted alongside coal possess a higher density than coal, typically with a particular gravity ranging from 1.35 to 1.5. Conversely, the refuse rock possesses a specific attraction that falls within the range of 2.1 to 2.3. Separation of these impurities from the coal can be achieved through a heavy-media process, whereby the ROM coal is submerged in a fluid that possesses a specific attraction bigger than the one of the original coal but lower than the fact of the impurity. Figure 3.3 illustrates how coal is made to float while heavy recyclable material is made to sink, thereby enabling the separate collection of both substances).

3.3.2.1. Dense-Media Washing

The dense-media procedure, or heavy-media, is widely employed to purify large-sized coal particles. On the other hand, Jig plants are considered the second most prevalent technique for treating coarse coal.

The cleaning process in a heavy-media scrubbing plant is accomplished exclusively through flotation, utilizing a medium of carefully chosen specific gravity. This medium is sustained by a dispersed form of magnetite that has been finely ground and mixed with water. The raw coal received is subjected to separation at a 1/4-inch interval on an inclined screen. The oversized coal particles are directed towards a level pre-wet screening surface, where the diminutive particulate matter is effectively dislodged from the coal that measures greater than one-fourth of an inch. The increase, as stated introduced through a vessel or bath containing a dense medium, whereby the waste material is segregated from the coal. The waste material is directed towards a screening mechanism known as a “refuse rinse” screen, which facilitates the removal of excess water. The utilization of magnetite has been examined for cyclone disinfecting of coal.

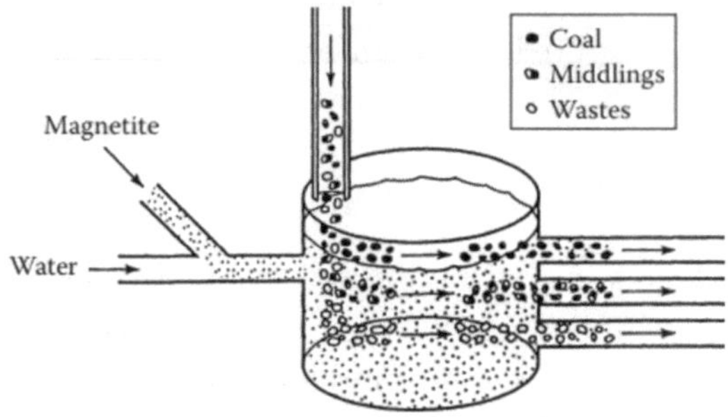

Figure 3.3. *Illustration of a dense-medium separation method.*

Source: James G. Speight, Creative Commons License.

The liberated medium is bifurcated into two segments, wherein one is reintegrated through the circulation pathway using the heavy-medium sump. At the same time, the other portion is conveyed to the magnetite recovery unit via pumping. The waste material is released from the mesh discarded. The coal undergoes release from the washer onto a coal-rinse television, which facilitates the dewatering process of the coal along with the treatment of the medium, similar to that of the trash screen. Subsequently, the clean coal undergoes a process of centrifugation, crushing, and loading. The coal particles smaller than 1/4 inch, as obtained through the initial coal screens, are mixed with water and magnetite. This mixture is then transported to a heavy-media ship, wherein the magnetite has a finer texture, and its particular gravity is altered.

The solid waste undergoes a process of dehydration, and the liquid component is retrieved, similar to the process employed during coarse coal choice. The coal undergoes discharge over a sieve bend, followed by subsequent transfer to a spin for ultimate dewatering. This is done before its transfer to either a heating dryer or a loading station.

The centrifugation process yields an effluent exhibiting a reduced presence of coal particles smaller than 28-mesh and separated from larger, uncontaminated coal fragments. The substance undergoes thickening through a cyclone, followed by de-sliming on a screen and centrifugation before being loaded. The utilization of heavy-media cyclones has become increasingly prevalent in processing fine-size fractions. Flotation is commonly employed to purify the 28-mesh size fractions. However, it has been observed that spirals alongside heavy-media hurricanes have also demonstrated efficacy in cleansing coal nourish down to 100-mesh. Spiral shapes are commonly employed for particle sizes ranging from – 10 to 60 mesh).

The utilization of sand processes involves the utilization of suspensions, which are frequently unstable, for sand within liquids. The efficacy of these processes relies on the maintenance of high anxiety levels and recirculation. It is possible to achieve specific gravity separations of up to 1.90; however, for bitumen coals, specific gravities ranging from 1.45 to 1.60 are typically utilized more frequently.

3.3.2.2. Pneumatic Cleaning

Pneumatic washing devices, also known as air tables, clean tiny portions that measure less than 3/8 inches. Within these apparatuses, air streams ascend via a perforated base plate beneath a coal stratum. The fine particulate matter is suspended in the atmosphere and necessitates recapture through the utilization of cyclones and bag filters without any enhancement in its quality. Upon nearing the terminus of these tables, the coal is stratified into distinct layers based on weight and ash content. The lowermost stratum comprises high-ash, weighty material, while the middle stratum consists of medium-weight fossil fuels and bone with high ash content. The uppermost stratum is composed of low-ash coal. The intermediate stratum must be integrated with the waste material (and subsequently cleansed) or coal.

The efficacy of these gadgets could be more optimal. The efficacy of their ash removal capacity is constrained to a range of 2% to 3%, irrespective of the quantity of ash present. These cleaning devices are characterized by their low expenditure on capital, lack of water supply, and disposal issues.

Screening of the arriving coal is a necessary step, and in certain plants, heated drying for the unprocessed feed is needed to ensure that the nourishment that enters the tables is devoid of moisture. Using heat dryers necessitates the implementation of cyclones and cleaners to regulate the release of particulate matter into the environment. Thermal dryers are commonly fueled by coal, oil, or gas.

3.3.2.3. Jig-Table Washing

The nomenclature "jig-table washing plants" is derived from using jigs for cleaning the >0.25 in. fraction and Diester tables for cleaning the <0.25 in. fraction. Diester charts are rotating sluices with a flat, riffled surface of about 12 ft^2 that swings parallel to the riffles and along the opposite direction toward the flow of fossil fuels. This kind of machinery can be used alongside froth cells and heat dryers.

The coal in its unprocessed state, limited to dimensions less than 8 inches, is subjected to wet screening using a mesh typically measuring 0.25 inches. The sizeable increment is introduced into the jig, while the residual coal is directed towards a distinct cleaning circuit. The coal undergoes a process of dehydration through the utilization of screens and centrifuges, subsequently followed by crushing to attain the intended particle size, culminating in loading. The jig employs getting in shifting currents to effectuate an "equivalent" gravity separation. Coal particles of diminutive size (measuring less than 1/4 inch) are mixed with an appropriate quantity of water and dispensed onto tables, where the undesired material is segregated from the coal. The solid waste is subjected to a dewatering process using a screen and subsequently disposed of. The dehydration of clean coal involves using a still gravity screen known as a sieve bend. This screen effectively removes the smallest particles, discharged through a centrifuge for ultimate dewatering. Subsequently, the clean coal with a size of +28 mesh is transported to a heating dryer through loading or conveying means. At the discharge point of the grid, the heavier rejects are released while, the lighter coal is released from the opposing end. The middlings are evenly distributed across the two ends.

3.3.2.4. Coal Drying

The moisture in coal has been found to affect its warming value adversely, handling characteristics, transportation expenses, and conversion produced during carbonation and other related processes. The decrease in the amount of water present is frequently a desirable outcome.

The presence of water within coal can be attributed to three distinct mechanisms: (i) inherent moisture that is confined within the internal pores within the coal material, inclusive of water that is associated with the substance contaminants; (ii) surface moisture that wets the outermost layers about the coal particles, wherein absorbing may play a minor role; and (iii) free water that is retained by capillary action in the interstices that connect the coal particulates.

The level of inherent humidity in coal is positively correlated with its rank, with the highest levels observed in low-rank lignites and brown charcoal. The presence of inherent humidity in coal can have a notable effect on the operational efficiency of low-rank coal, which is predominantly utilized for power generation. Typically, the process of drying is limited to two categories of coal: (i) washed bitumen coal that must adhere to specific user requirements along with (ii) lignite or sub-bituminous fossil fuels that are utilized in the production of briquettes or other specialized items, including absorbing carbon.

The surface humidity in coal is influenced by the extent of the surface that is accessible and the coal's wetting. However, it is noteworthy that the moisture levels are comparatively lower than anticipated, given the available surface area, owing to coal's relatively low wetness to mineral surfaces. Surface humidity from higher-grade coal varieties such as bitumen and anthrax can be eliminated through screen drainage. In contrast, the dehydration of smaller coal particles or coal fines is possible through cyclones and centrifuges. In instances where it is necessary to decrease the amount of moisture to achieve lower levels, two potential techniques involve the utilization of either rotary kilns or liquid bed dryers. However, it should be noted that dried coal has the potential to reabsorb moisture upon being exposed to the environment, thereby posing a risk of fire and explosion hazards.

Thermal dripping has been widely utilized for metallurgical burning coal in certain regions of the coal industry, and there is an increasing desire to establish effective techniques for dripping low-rank coals. Thermal drying involves the exposure of moist coal particles to elevated temperatures through hot gases, typically derived from combustion processes. This method facilitates the evaporated water of surface moisture from the coal particles while avoiding any potential damage or burning of the coal.

During operations, the clean coal from diverse wet cleaning techniques exhibits a moist consistency, necessitating a drying procedure to render it appropriate for conveyance and ultimate utilization. The thermal drying process is utilized to dry coal with high moisture content.

3.4. COAL SAMPLING

The act of collecting coal holds significant value in the realm of process control within a coal preparation facility. A grasp sample refers to a singular portion of coal extracted from a specific point within the process stream, which may not accurately reflect the overall characteristics of the coal. A regular specimen is collected at a predetermined interval, during a specific duration, or for each consignment. The coal sampling process involves using various sampling instruments, including the crosscut sampler, which emulates the "stop belt" sample and determines the appropriate approach to coal testing. The cross-cut samples are positioned presently on the upper surface of the conveyance belt. The slipping stream samples are placed at the upstream section of the conveyor belt.

Coal operations typically select numerous locations within the wash plant to extract samples of unprocessed coal. The three samples that are typically analyzed in a coal processing plant include (i) the coal-related decline entering the plant; (ii) the raw coal rejected to assess its characteristics; and (iii) the scrubbed coal to evaluate its quality for shipment.

After the collection of gross samples, it undergoes a crushing process and is subsequently quartered to get a net sample. This net sample is then forwarded to an autonomous laboratory for evaluation, and the outcomes are communicated to the purchaser and the provider. Frequently, the purchaser may solicit a duplicate or a subsequent examination from alternate testing to ensure the veracity of the information. The coal's ash, humidity, heat content (expressed in Btu/lb), sulfur calcium, magnesium, iron sodium, and various other fundamental components are continuously measured and subsequently reported.

3.5. TRANSPORTATION

Coal is frequently transported via rail, road, and water from the mining site to the market. In certain coastlines, coal was transported via conveyors compared to the mines to the cargo holds of sizable maritime vessels. In Britain, a significant portion of the coal extracted through the north coalfields is transported to the southern regions via colliers, maritime vessels specifically designed for this purpose. Typically, the transportation of coal via truck on a large scale is deemed financially feasible solely for short distances.

Over the past few years, a trend has emerged wherein significant industrial facilities, such as those involved in chemical-based or power production, are erected near mining sites. This is done to minimize coal transportation expenses, and the coal is transported directly to the facility through high-capacity trucks or conveyor belts. The concept of transporting oil sand via a conveyor system over long distances is employed by oil sand processing facilities located in northern Alberta. The sand is transported for several miles before reaching the manufacturing plant.

Transporting synthetic petroleum to the marketplace is more feasible and cost-effective than transporting low-value oil sand. In comparison, the cost of transmitting electricity over extended distances through high-voltage wires is significantly lower than the expense of transporting the same amount of coal. Nevertheless, the feasibility of implementing this idea could be hindered by the substantial expenses of building a facility in a geographically isolated location with an inhospitable climate near a mining site. The proposal to transport coal from the mining location using coal/water and coal/oil slurry within pipelines is a potentially advantageous solution. This method could provide a convenient and efficient means of transporting coal to markets by utilizing pre-existing pipeline infrastructure).

3.5.1. Unit Train

The predominant mode of transporting coal over extended distances is using unit trains. A unit train is a collection of railcars utilized for a specialized shuttle service between a coal mine or a power plant.

A standard unit train comprises 10 to 120 railcars and 3 to 5 locomotives, whereby each railcar has a capacity of approximately 100 metric tons (105 kilograms) of coal-fired power. Efficient management of the loading and unloading of terminals is imperative to minimize expenses. A unit train that completes a round trip from the mine to the plant is characterized by a standard turnaround duration of 72 hours, encompassing a 4-hour loading period and a 10-hour unloading and servicing interval per train.

The system has been devised to enable the loading and unloading of trains requiring them to come to a halt, thereby facilitating uninterrupted transportation of coal and augmenting the speed of coal movement from the mining location to the end user. On numerous occasions, it proves to be a challenging task to navigate through various roads in the eastern region of Wyoming due to the frequent occurrence of coal-carrying trains en route to their destination. Rail transportation serves as the primary means of transportation for the vast majority (95%) of western coal mining operations, spanning from the mountainous areas of Wyoming to the northern regions of Alaska.

3.5.2. Coal Barge

Barges have a significant role in the transportation of coal in the USA and Europe, particularly on rivers and lakes. Barges that transport coal are typically organized into tows consisting of 15 to 40 barges propelled by just one towboat with a horsepower ranging from 2,000 to 10,000. A barge of "jumbo" size can transport 1,800 tons of coal. Therefore, a substantial tow has the potential to convey 72,000 tons of fossil fuels, which is equivalent to the amount transported by five-unit trains. The substantial magnitudes of such volumes lead to noteworthy economics of scale and reduced expenses. The price of barge transportation, calculated per-mile and per-kilometer basis, could be as low as 25% or less of the expense of rail transportation. Water transportation routes frequently exhibit meandering paths, leading to prolonged transit durations.

3.5.3. Coal Pipeline

A slurry pipeline is an alternative means of coal transportation. This infrastructure facilitates the transportation of coal from a mining site to a power generation facility, where it is utilized as a fuel source for producing electrical energy. Coal slurry pipes transport pulverized coal and water, commonly called slurry. The coal-to-water proportion is approximately 1:1.

The coal extracted from the mine undergoes a crushing process to attain a diameter of approximately 1 mm. Subsequently, it is combined with water in storage containers equipped with agitators, which facilitate the maintenance of the coal within suspension within the water. The annual water consumption of the pipeline amounts to approximately 1 billion gallons. After three days, the slurry is transported through the pipeline to its final destination at the Mohave nuclear facility. Upon arrival, it is stored in anxious tanks over immediate utilization and in drying ponds for future use. Heated centrifugation is employed for dehydration purposes. In 2006, the plant ceased operations due to ongoing negotiations regarding water and coal availability terms.

Coal slurry pipes have been identified as a possible cost-effective mode of coal transportation with favorable economic implications. The veracity of this claim about a specific pipeline can solely be ascertained through a comprehensive assessment of the route's conditions. The current coal transportation situation lacks options among slurry pipelines along with railroads, which would inevitably reduce the cost of coal transportation. Within the current context, evaluating the potential financial, ecological, and social ramifications of fossil fuels slurry pipeline construction and the distribution of coal via this method is imperative.

Nevertheless, it is imperative to exercise prudence. The considerable water and energy demands associated with coal slurry represent a notable impediment to its expanded implementation, particularly in dry areas of Australia and the western portion of the United States.

The charcoal log pipeline is a technological innovation that involves the compression and shaping of coal into cylindrical forms (coal logs) at the mining location for transportation. Subsequently, the coal accesses are introduced into a subterranean conduit that contains water for conveyance to a designated endpoint. This endpoint may comprise singular or multiple power plants, a railway station, a dock for barges, or a harbor for intermodal conveyance.

3.5.4. Truck Transport

The transportation of coal is commonly facilitated by end-dump vehicles, which possess a weight capacity of approximately 25–50 tons. Truck delivery is widely employed for small-scale power plants in the United States' eastern region.

Coal transportation can be facilitated through the use of trucks that are capable of carrying 15–30 tons of coal and can travel on standard highways. Coal can be conveyed through oversized off-highway vehicles with a load-carrying capacity between 100 and 200 net metric tons. Typically, these vehicles are equipped with diesel engines and feature either back or bottom disposal of mechanisms.

Roads specifically designed for transporting coal are widely used for power plants near coal mines in the western, southern, and eastern regions. On the other hand, the transportation of coal through trucks on roadways is more prevalent in surface mines. In the United States, the typical range for truck pulls on public roads is approximately 50–75 miles, whereas off-road pulls typically span 5–20 miles.

3.5.5. Ocean Transport

The Richards Bay Mining Terminal in South Africa is recognized as the largest coal-export facility globally. The volume of coal traded globally experienced a notable rise, ascending from 498 Mt in 1990 to 917 Mt in 2007. It is anticipated that there will be a continued increase in coal exports, resulting in a corresponding elevation in the significance of maritime transportation. The shipment of coal via ocean necessitates a comprehensive framework comprising several components, including the shipment of coal from its source to the harbor, the availability of coal-handling places at the destination port, the existence of ocean carrier systems with sufficient numbers and sizes of ships, the fulfillment of contractual responsibilities, the effective management of the maritime fleet, and informed route selections. Additionally, coal-handling establishments at the import interface and the journey from the entry point to the customer are crucial aspects of this process.

Small fishing boats to massive cargo vessels.

i. The vessel is of a convenient dimension, with a dead weight tonnage (DWT) ranging from 40,000 to 45,000. The term typically refers to watercraft with a deadweight tonnage ranging from approximately 10,000 to 40,000.

ii. The Panamax class of vessels has an estimated DWT range of 60,000 to 80,000. From a technical standpoint, the Panama Canal has a restriction of 32.2 meters in beam, which limits the optimum size of vessels that can travel through it.

iii. The category of ships referred to as Cape-size vessels, with a DWT exceeding 80,000, is the subject of discussion. A maritime vessel of dimensions exceeding the maximum allowable size for transit through the Panama Canal, necessitating a route via the Cape for Good Hope for passage between the Pacific and Atlantic Oceans, and vice versa.

3.5.6. Conveyer Belt Transport

Conveyors are commonly employed in power plants located at the mine's entrance to facilitate the transportation of coal compared to the mining site to the preservation or utilization zone. Conveyor belts are a viable option for the transportation of coal in regions with rugged topography where conventional roadways are not easily accessible. These belts are commonly utilized for conveying coal over distances ranging from 5 to 15 miles.

Conveyors possess the benefit of requiring minimal maintenance. However, they need more location flexibility, necessitating a truck for transportation. Mobile conveyor belts are being developed and utilized. The singular negative environmental ramifications associated with conveyor belts utilized for the transportation of coal are limited to the loss of coal dust that may occur during loading, removal, or transportation processes.

3.6. STORAGE

The storage for coal is a crucial component of coal-handling structures at mining sites, especially in light of the unit-train approach to transportation.

Coal storage is typically employed to achieve one or more of the goals that follow: (i) to ensure prompt readiness for transportation; (ii) to enable mixing both the physical and chemical discrepancies inherent in this diverse material, as well as a uniform, marketable product; and (iii) to keep coal of specific sizes in response to seasonal demand.

Coal storage is deemed essential for both the mining and utilization aspects. Nonetheless, keeping of coal might be supported by several unfavorable occurrences, among which the most significant are (i) oxidization and accidental combustion; (ii) alterations in coal characteristics that could impact its utilization; (iii) deterioration of coal because of multiple handling procedures; and (iv) the expenses associated with the multi-handling procedures and storage facilities.

Nevertheless, coal reacts with surrounding oxygen, even under normal conditions, releasing heat. Suppose the exothermic heat generated throughout the process is permitted to build up. In that case, the speed of the preceding reaction experiences an exponential increase, and there is a subsequent elevation in temperature. Upon reaching the point of ignition of coal, a phenomenon known as spontaneous ignition occurs, wherein a chemical reaction that is exothermic inside the combustible material leads to a subsequent increase in temperature and results in spontaneous combustion.

Did You Know?

The world's largest coal storage facility is located in Northumberland, England known as the Blyth power station coal yard.

The self-sustaining oxidation reaction of coal and subsequent spontaneous combustion is contingent upon its type, encompassing its nature, rank, and the dissipation or absence of heat. The degree of Celsius at which this reaction occurs is subject to variation. In cases where coal is of inferior quality and exhibits high heat retention, combustion can commence at temperatures in the range of 30°C–40°C (86°F–104°F).

3.6.1. Oxidation and Spontaneous Combustion

The propensity of coal to undergo spontaneous heating during storage is primarily contingent upon its susceptibility to oxidation. The tendency of coal to become oxidized is closely associated with various factors, including coal position, size uniformity or arrangement of coal within the pile, stockpiling method, stockpiling temperatures, the pyrite material, moisture level, air circulation circumstances, storage duration, and the presence of outside supplies. Coal rank exhibits an inverse relationship with its propensity to oxidize, with greater ranks exhibiting lower tendencies. Smaller pieces of coal possess more surface area that allows oxygen to interact with it, which increases the likelihood of oxidation. The approach and temperature utilized for stockpiling and the quantity and form of pyrite present also affect the tendency of coal to oxidize. Moisture concentration and circulation circumstances throughout the pile, as well as the involvement of foreign materials, are additional factors that contribute to the inclination of coal to oxidize. Furthermore, the natural variability of coal, coupled with the factors above, precludes precise prognostication of the onset of unplanned ignition (spontaneous combustion).

3.6.1.1. Oxidation

The exothermic nature of oxidation results in the generation of heat at greater speeds than can be dissipated and expelled from the accumulated by air circulation, owing to the increase within the rate of a chemical reaction by 10°C (18°F). Consequently, the temperature will increase to a level where self-sustained combustion is initiated and sustained.

Problems may arise when dealing with fossil fuels with high levels of moisture and sulfur, as well as coal which has a propensity to deteriorate upon exposure to atmospheric oxygen. The matter is of utmost importance when considering low-rank and high-sulfur-content coals. The storage of lignite and coal poses a challenge due to the risk of explosive combustion, unlike anthracite, which has a significantly lower likelihood of experiencing spontaneous igniting.

3.6.1.2. Oxidation and Rank

The capacity of coal to absorb oxygen, leading to weathering or combustion, exhibits significant variability. Typically, oxygen absorption diminishes with an increase in rank, with anthrax demonstrating low absorption along with coal and lignite exhibiting high intake. Oxygen absorption is comparatively greater in coals with elevated bed moisture, which is the natural moisture present in the coal bed. This is determined by various factors such as capacity moisture, organic bed humidity, a state of equilibrium humidity (ASTM,

2011a), oxygen level, and uncertain content. Along with high sleeping moisture, such coals are usually classified as low-rank charcoal.

The occurrence of natural combustion is an occurrence that is closely associated with the level of rank. The proclivity of coal towards self-heating exhibits a negative correlation with its rank, wherein lignite and sub-bituminous burning coal demonstrate a greater vulnerability to heating itself bituminous coal along with anthracite. A reduction in rank is associated with increased inherent humidity, volatile issue, and hydrogen and oxygen. Coal samples with medium to high volatility and a volatile material content exceeding 18% w/w daf exhibit a higher oxidation rate than low-volatile coal samples. Consequently, they are more susceptible to burning spontaneously than their low-volatile counterparts. Lower-rank coals exhibit higher porosity than their higher-rank counterparts, resulting in a larger surface area susceptible to oxidation. Coals with low rank possess an extended chain of hydrocarbons, thereby exhibiting lower stability than anthracite, which has a relatively reduced hydrocarbon constituent. Nevertheless, there can be a considerable range of variation in the oxidation rate among coals of identical rank.

The oxidation process of coal is commonly acknowledged to occur in five distinct stages, primarily influenced by the temperature and chemical reactions involved. However, it should be noted that this consensus is not entirely universal:

i. Slow oxidation initiates in coal until it attains a temperature of around 50°C (122°F).
ii. At this juncture, the oxidation process experiences an escalation in its rate till the coal's surface temperature reaches around 100°C–140°C (212°F–285°F).
iii. At around 140°C (285°F), the coal undergoes a process whereby water and carbon dioxide water vapor are generated and discharged.
iv. The release of carbon dioxide experiences a rapid increase as the temperature reaches 230°C (445°F), potentially leading to unplanned ignition and combustion.
v. At a temperature of 350°C (660°F), coal undergoes ignition and experiences a robust combustion process.

Typically, the critical range of temperatures for bitumen coal during storage is around 50°C–66°C (122°F–150°F). Upon reaching this particular temperature, the heating process typically experiences a rapid escalation and may become unstable, ultimately leading to ignition, unless preemptive measures are implemented.

At a consistent level of rank, it has been observed that the heating itself tendency of coal reduces with an increase in its inertinite material. The overall pattern also suggests a rise in the tendency for self-heating as the vitrinite and liptinite material increases.

Liptinite > Vitrinite >>Inertinite

The self-heating phenomenon in coal appears to correlate more closely with coal rank than its petrographic structure.

3.6.1.3. Pyrite

Pyrite is various minerals contain chemicals that can impact the oxidation rate by hastening or impeding the process. Alkalis have the potential to accelerate the process, whereas borates, along with calcium chloride, can serve as retardants. The presence of ankerite within the mineral substance found in coal can facilitate the procedure of oxidation. On the contrary, the oxidation process is slowed down by the presence of silica along with alumina.

Pyrite, a mineral of iron disulfide (FeS_2), generates thermal energy through aerial oxidation and has been hypothesized to be responsible for coal self-heating. The process of self-heating in coal is facilitated by localized heat generation. However, the reaction products from this process exhibit more volume than the initially formed pyrite. Consequently, any coal that contains these products is prone to breaking open, thereby exposing a larger surface area of coal to the surrounding air.

The exothermic reaction that occurs between pyrite (FeS_2) and oxygen, as well as water, leads to the production of both iron sulfate ($FeSO_4$) as well as sulfuric acid (H_2SO_4). It can be inferred that the exposure of coal to rain may accelerate the reaction rate, and similarly, water-flooding as a means of extinguishing fires might have the same influence on the reaction rate.

3.6.1.4. Coal Size and Stockpile Ventilation

The oxidation process is positively correlated with the degree of coal fineness, whereby finer coal particles exhibit a greater propensity for oxidation. Additionally, the rate of coal oxidation in the presence of atmospheric oxygen is directly related to the particular internal surface area of the coal.

The empirical findings indicate that the corresponding low-temperature coefficients follow a cubic root function. Additionally, the investigation reveals that the rate and magnitude of oxidation exhibit an upward trend as the particle size decreases until a critical diameter is attained. Subsequently, the rate of oxidization stabilizes at a relatively constant level.

Typically, the inherent airflow in coal preservation piles is sufficient to expel sensible heat at a rate equivalent to its release during the oxidation phenomenon. In scenarios where the airflow is sufficient to sustain oxidation yet insufficient to disperse the heat generated, the coal assimilates the heat, increasing its internal temperature. A cascade of events occurs whereby the oxidation rate escalates with rising temperatures. If left unmitigated within a storage heap, the rock's ignition temperatures will ultimately be attained, initiating a smoldering process.

The storage of ROM coal poses a challenge due to the significant presence of fines in conjunction with the lump. Conversely, the risk of storing double-screened and strongly calibrated lump coals is minimal. The coal pieces exhibit a uniform structure characterized by a honeycomb-like network of passages, facilitating the free circulation of air. This, in turn, enables the efficient dissipation of the heat generated. The risk of unplanned ignition can be reduced by compacting coal fines reserves to exclude air.

3.6.1.5. Moisture Content

The impact of moisture on natural combustion in coal stockpiles is well-documented, as it can significantly alter ventilation and pyrite reaction.

The impact of moisture upon self-ignition is dual. Specifically, the procedure of moisture vaporization results in energy consumption, thereby hindering the ignition process. Conversely, there has been an observation of an increase in self-ignition through soaking materials susceptible to this phenomenon.

Dry coal + Moisture Wet coal + Heat

When present, moisture generates heat due to wetting and obstructs oxygen ingress into the coal's pores. The outward diffusion of water vapor via pores reduces oxygen relative pressure, thereby causing a decrease in the reaction rate. Alternatively, polarized may bind to reactive sites within the coal.

3.6.1.6. Time Factor

Oxidation occurs upon the exposure of a newly subjected coal surface to air. Nevertheless, the oxygen absorption rate has become inversely related to time as the ambient temperature stays constant. Hence, it can be inferred that stockpiling coal in a manner that prevents a significant increase in temperature within the pile. By ensuring that heat is dissipated at a rate equal to or greater than the heat generation rate resulting from the oxidation procedure, the rates of oxidation and consequent degradation or weathering caused by the coal will decrease over time.

3.6.2. Effect of Storage on Coal Properties

Coal kept in storage may experience a decline in its heating value along with coking quality. Typically, superior quality coal (stored appropriately to restrict oxidization to the lowest) experiences a mere 1% annual reduction in its calorific value. Conversely, inadequate storage practices may lead to a reduction in heat value of approximately 3% to 5% within the initial year.

Furthermore, the coking properties of numerous coal types and coal mixtures are significantly impacted by aging during storage, rendering them unsuitable for utilization as a charge in coke ovens. The available data and assertions exhibit a range of outcomes, from (i) absence of impact on the combustion properties following prolonged storage periods to (ii) notable deterioration of coking properties within a mere month of storage. The long-term preservation of low-rank coal poses specific challenges due to its tendency to experience reduced strength, deterioration, and decreased heating value.

3.6.3. Elements of Safe Mine Storage and Reclaiming of Coal

3.6.3.1. Long-Term Storage

When coal is intended to be stored for prolonged durations, such as throughout a winter season or when sales are low, it is recommended that the storage location(s) be devoid of moisture and possess adequate drainage capabilities. This can be achieved through natural topography or by implementing appropriate construction techniques. It is necessary to eliminate all flammable materials with low temperatures, including rags, wood, evaporated hay, and other comparable items, from the designated area.

For constructing a preservation pile, it is recommended to use clay or densely packed earth as the foundation. The coal ought to be evenly distributed over the entire surface with a thickness of around 1–2 feet and then compressed. It is recommended to refrain from creating conical heaps and, instead, compact and roll both sides and the top of every pile to establish a seal and prevent air infiltration. A viable solution for sealing a coal pile involves utilizing a seamless layer of fine coal, which is subsequently overlaid with lump coal to mitigate the risk of seal erosion due to the influence of wind and precipitation.

The storage of examined coal in larger sizes poses minimal challenges. Typically, a ventilation system that permits natural airflow to dissipate the minimal amount of heat generated is sufficient for storage purposes. Furthermore, compressed fine coal seals can be utilized. The mitigation of fines production can be achieved in cases where coal exhibits a propensity to slack. The recommended storage approach for ROM coal and stoker-sized fossil fuels is the layering technique, which involves sloping edges for drainage purposes. The application of oil treatment upon smaller coal sizes during storage is occasionally preferred due to its ability to impede the uptake of water and air.

3.6.3.2. Short-Term Storage

The recommended measures for the extended preservation of coal can be extended to the short-term keeping of coal in reserves, especially at loading facilities that handle unit trains and have reclaiming tunnels. The primary risk linked with coal recuperation tunnels pertains to the potential development of a combustible environment that arises from coal dust and methane buildup. The build-up of methane can occur despite seemingly sufficient ventilation practices, while the dust buildup is contingent upon the level of surface moisture in the coal.

The emission of propane to coal can result in a stagnant atmosphere and potentially impede low-temperature oxidation, particularly in coals with a high gas content. However, it should be noted that methane also holds promise as a viable energy source. In addition, it should be noted that a significant increase in the amount of coal surface area susceptible to oxidation accompanies the reduction in methane in the desorption process.

3.7. SUMMARY

Coal is a fossil fuel that has been widely used for centuries as a source of energy. The preparation, transportation, and storage are important aspects of its utilization. Coal is typically extracted from underground mines or open-pit mines. After extraction coal is crushed to reduce its size and make it suitable for transportation via rail, barge, or truck. Coal is often stored in large outdoor piles or stockpiles. In some cases, coal is stored in enclosed structures such as silos or bunkers. Power plants and industrial facilities may have dedicated coal yards where coal is stored. It's worth noting that the preparation, transportation, and storage of coal can have environmental impacts, including air and water pollution.

REFERENCES

1. Agrawal, P. L., (1959). *Proceedings of the Symposium on the Nature of Coal* (p. 121). Central Fuel Research Institute, Jealgora, India.
2. Baughman, G. L., (1978). *Synthetic Fuels Data Handbook*. Cameron Engineers, Denver, CO.
3. Berkowitz, N., (1979). *An Introduction to Coal Technology*. Academic Press, San Diego, CA.
4. Braunstein, H. M., Copenhauer, E. D., & Pfuderer, H. A., (1977). *Environmental Health and Control Aspects of Coal Conversion: An Information Review*. Report ORNL-EIS-94. Oak Ridge National Laboratory, Oak Ridge, TN.
5. Chatterjee, I., & Misra, M., (1989). *Proc. Mater. Res. Symp. Mater. Res. Soc., 189*, 195.
6. Davis, A., (1978). In: Karr, Jr. C., (ed.), *Analytical Methods for Coal and Coal Products* (Vol. 1). Academic Press, San Diego, CA, Chapter 2.
7. Davis, A., Mitchell, G. D., Derbyshire, F. J., Rathbone, R. F., & Lin, R., (1991). *Fuel, 70*, 352.
8. Dorman, H. N. M., Huntjens, F. J., & van Krevelen, D. W., (1957). *Fuel, 36*, 321.
9. Ergun, S., (1951). *Anal. Chem., 23*, 151.
10. Ettinger, I. L., & Zhupakhina, E. S., (1960). *Fuel, 39*, 387.
11. Evans, D. G., & Allardice, D. J., (1978). In: Karr, Jr. C., (ed.), *Analytical Methods for Coal and Coal Products* (Vol. 1). Academic Press, San Diego, CA, Chap. 3.
12. Gan, H., Nandi, S. P., & Walker, P. L., Jr., (1972). *Fuel, 51*, 272.
13. Kalliat, M., Kwak, C. Y., & Schmidt, P. W., (1981). In: Blaustein, B. D., Bockrath, B. C., & Friedman, S., (eds.), *New Approaches in Coal Chemistry*. American Chemical Society, Washington, DC, Chapter 1.
14. King, J. G., & Wilkins, E. T., (1944). *Proceedings of the Conference on the Ultrafine Structure of Coal and Cokes* (p. 45). British Coal Utilization Research Association, London.
15. Kroger, C., & Badenecker, J., (1957). *Brennst. Chem., 38*, 82.
16. Labuschagne, B. C. J., (1987). *Proceedings of the 4th Annual Pittsburgh Coal Conference* (p. 417). University of Pittsburgh, Pittsburgh, PA.
17. Labuschagne, B. C. J., Wheelock, T. D., Guo, R. K., David, H. T., & Markuszewski, R., (1988). *Proceedings of the 5th Annual International*

Pittsburgh Coal Conference (p. 417). University of Pittsburgh, Pittsburgh, PA.

18. Lyman, W. J., Reehl, W. F., & Rosenblatt, D. H., (1990). *Handbook of Chemical Property Estimation Methods: Environmental Behavior of Organic Compounds*. McGraw-Hill, New York.
19. Mahajan, O. P., & Walker, P. L., Jr., (1978). In: Karr, Jr. C., (ed.), *Analytical Methods for Coal and Coal Products* (Vol. 1). Academic Press, San Diego, CA, Chapter 4.
20. Milewska-Duda, J., (1991). *Arch. Mineral. Sci., 36*, 369.
21. Ramesh, R., Francois, M., Somasundaran, P., & Cases, J. M., (1992). *Energy Fuels, 6*, 239.
22. Speight, J. G., (1978). In: Karr, Jr. C., (ed.), *Analytical Methods for Coal and Coal Products* (Vol. 2). Academic Press, San Diego, CA, Chapter 22.
23. Speight, J. G., (1994). *The Chemistry and Technology of Coal* (2nd edn). Marcel Dekker, New York.
24. Speight, J. G., (1999). *The Chemistry and Technology of Petroleum* (3rd edn.). Marcel Dekker, New York.
25. Van Krevelen, D. W., (1957). *Coal: Aspects of Coal Constitution*. Elsevier, Amsterdam.
26. Walker, P. L. Jr., (1981). *Philos. Trans. R. Soc. (London), 300A*, 65.

CHAPTER

4 Coal Analysis

Contents

INTRODUCTION

Coal is a naturally occurring organic sedimentary stone that comprises diverse proportions of nitrogen, oxygen, carbon, hydrogen, and sulfur alongside insignificant quantities of other elements, such as mineral matter (Donahue & Rais, 2009).

It is believed that the term "coal" originated from the Ancient English word "col," which referred to a specific charcoal used during that era. Coal is commonly known as sea coal in certain regions due to its occasional appearance on beaches, particularly in northeastern England. In the early Middle Ages, coal mining was not a common practice (Zhu, 2014). However, written records indicate that coal mining began to occur after the year 1000. The utilization of coal experienced a significant surge during the 19th and early 20th centuries. The surge in demand for coal has necessitated the development of appropriate techniques for coal analysis to establish a relationship between fuel characteristics and performance.

Coal is a carbonaceous rock characterized by its solid, brittle, and combustible nature (Potgieter-Vermaak et al., 2011). It is formed through the decomposition and transformation of vegetation, subjected to compacting, temperature, and pressure. The object's hue ranges from brown to black and typically appears stratified. Vegetation, in many cases, is derived from moss and other types of low-lying plant species. However, certain coal deposits exhibit considerable substances that can be traced back to woody antecedents.

Over a prolonged geological period, the plant precursors underwent compaction, chemical alteration, metamorphosis, and hardening due to heat and pressure, ultimately forming coal. The prevailing hypothesis is that coal originated from ancient flora that thrived in wetland environments (Sheta et al., 2019). Upon the death of these plants, their dead tissue became embedded in anaerobic aquatic environments, where the low levels of oxygen impeded their decomposition, thereby preventing the release of carbon dioxide. The accumulation of unoxidized organic material resulting from the growth and decay of subsequent generations of this plant type led to carboniferous deposits, including peat, bituminous, or anthracite coal, after being buried by sediments and compressed. Occasionally, shale, and sandstone sediments above coal deposits can provide evidence of the plant species that formed carboniferous deposits.

Coal deposits, commonly called beds or seams, exhibit a wide range of thicknesses, spanning just a fraction of an inch to several 100 feet. Coal deposits can be traced back to geologic periods ranging from Silurian to Quaternary (Ward, 2016). However, the initial coals that held significant commercial value were discovered in the Mississippian rock formations (The carboniferous era in Europe). Coal deposits are typically generated in fluvial environments or basins exposed to marine incursions. Coal deposits are distributed across all continents, and the global reserves of coal surpass the 1 trillion-ton mark.

Coal is composed of over 50% by weight and over 70% by carbonaceous material volume, encompassing inherent moisture (Ward, 2002). The primary application of this substance is as a solid fuel to generate heat through combustion, resulting in the emission of carbon dioxide, a potent greenhouse gas, and sulfur dioxide. The resultant outcome is the generation of sulfuric acid, which is accountable for the creation of sulfate aerosol and the occurrence of acid rain. Coal is known to possess numerous trace elements, such as arsenic and mercury, that can pose a significant threat to the environment if they are discharged. Coal possesses trace amounts of thorium, uranium, and various other naturally existing radioactive isotopes, which may result in radioactive contamination upon being released into the environment. Despite being present in small quantities, the combustion of a substantial amount of coal results in the emission of considerable quantities of said substances. The process of utilizing coal for electricity production involves the conversion of heat energy into steam, which subsequently drives the turbine generators. Coal significantly contributes to global electricity production, accounting for approximately 40% of the current output (Kiss & King, 1977). The recoverable coal deposits, utilizing present-day technologies, are estimated to be adequate for a minimum of 300 years of consumption. Contemporary coal-fired power plants employ diverse methods to mitigate the harmful effects of their byproducts and enhance combustion efficiency. However, adopting these methods is not widespread in certain nations due to their augmentation of the power plant's initial investment. Bituminous coal is a highly compact coal typically characterized by its black or dark brown coloration and frequently exhibits distinct bands of contrasting brightness and dullness. Its primary application is as a fuel source for generating steam-electric power (Balachandran, 2014). However, it is also utilized in significant quantities for heat and power purposes in manufacturing and in the production of coke. Typically, the weight percentage of moisture in bituminous coal is below 20%. The heat capacity of bituminous coal varies between 21 to 30 million Btu per ton, based on a moist, mineral-matter-free foundation (Figure 4.1).

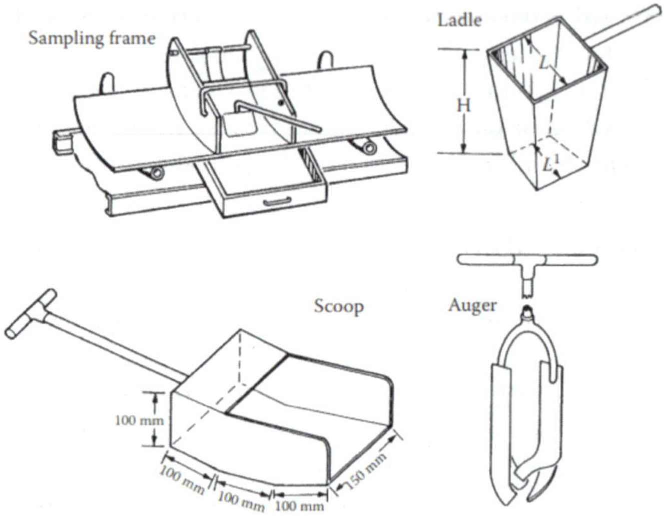

Figure 4.1. *Illustration of the sampling tools.*

Source: Meyers, Creative Commons License.

Remember

Coal samples used for analysis are typically collected from coal mines, coal seams or coal ship

4.1. ANALYSIS CONSIDERATIONS

The coal analysis data determines the cost of coal production by allocating production costs, regulating mining, and cleaning operations, and evaluating plant efficiency (Warwick, 2005). It is imperative to acknowledge the constraints of the analytical techniques employed. In the context of commercial operations, the price of coal is determined not solely by the amount of coal but also by the correlation between a desirable characteristic or a combination of characteristics and the coal's performance in service conditions.

The quantification of the intended characteristics, commonly called specifications, is articulated numerically. Hence, the precision of these quantifications holds significant value. The measurements must be precise enough to avoid adverse scientific or economic outcomes. The numerical data obtained from experimental procedures must conform to the established error limits of the test methods employed (Eskanazy et al., 2010). This ensures that the data can be considered precise absolute values rather than mere approximations. The utilization of statistical analysis in such testing methodologies necessitates a highly cautious approach. The analysis in question must rely on sound assumptions and avoid any potential accusations of mathematical manipulation aimed at producing a desired outcome. Stated differently, utilizing dependable standardized testing techniques to conduct coal analysis is necessary.

The analysis of coal poses several challenges, primarily due to its heterogeneous nature (Sheta et al., 2019). Additional issues encompass the propensity of coal to absorb or discharge moisture and to experience oxidation upon exposure to the ambient environment. Moreover, the substantial examinations and evaluations necessary to appropriately characterize coal also raise concerns.

According to its definition, a standard is a written document developed through a consensus process and ratified by an authoritative entity. Its purpose is to furnish rules, guidelines, or features for activities or outcomes intended for frequent and widespread application. Numerous industry organizations and trade associations must adhere to a specific standard or directive for a product (such as coal) before its commercial availability (Zhao et al., 2021). The utilization of standards is increasingly becoming a fundamental requirement for global commerce. Primarily, adherence to standards can provide any business, regardless of size, with advantages in terms of consistency and ethical conduct.

The establishment of diverse national standards associations has resulted in the emergence of techniques for assessing coal. The American Society for Testing and Materials (ASTM) has continuously worked in this area for numerous decades (Żogała, 2014). At the same time, research about establishing standardized techniques for coal assessment has been conducted in all major coal-producing nations. Apart from the ASTM, national-level organizations are dedicated to developing and standardizing methods. The British Standards Institution (BS) and the International Organization for Standardization (ISO) are two examples. The former covers coal analysis under a single standard number.

The post-World War II surge in trade among coal-producing nations necessitated cross-referencing established standards, which was entrusted to the ISO based in Geneva, Switzerland. Participation in this organization is granted to actively involved countries or holding observer status (Han et al., 1978). In addition, novel techniques are consistently emerging within the comprehensive scheme of coal assessment. At the same time, the established methods may require periodic adjustments to enhance the preciseness of the outcomes and the reliability of the approach.

It is pertinent to acknowledge the relevance of the corresponding test when discussing the specific techniques employed in the assessment of coal for coal-based commodities. The essential ASTM, BS, and ISO test numbers have been incorporated as appropriate.

A modified version of a proximate or ultimate analysis is frequently solicited in conjunction with any or all of the diverse analyses or examinations deliberated in this section (Dooher, 2005). The imposition of limitations on the coal utilized in coal-fired power plants and other coal-burning facilities has generated a demand for enhanced coal assessments and more precise and expeditious analytical techniques. This trend is expected to persist, necessitating further testing as the utilization of coal in the liquefaction along with gasification facilities continues to expand.

4.2. ACCURACY AND PRECISION

The attainment of accuracy and precision is imperative in any analytical endeavor, as the reliability of the resulting data is compromised without them, rendering it unreliable for practical use (Hou et al., 2016). This holds particularly true for analytical data employed in commercial operations, wherein the merchandise is marketed based on its purity level. Coal is a material of intricate composition, prompting inquiries into its purity. In this context, purity pertains to the presence or absence of extraneous elements inside the organic coal structure. The foreign constituents in the substance under consideration include water, pyrite, and minerals. Hence, it is recommended to acknowledge the inherent distinctions between the concepts of accuracy and precision.

The term "accuracy" denotes the dependability of an assessment or observation (Bona & Andrés, 2007). However, it pertains more precisely to the degree of proximity between an experimental outcome and the actual value. The level of concurrence between individual test outcomes and a recognized benchmark value determines the precision of a given test methodology. In contrast, precision can be defined as the extent to which duplicate data and measurements adhere to one another and the level of concurrence among individual test outcomes acquired under comparable predetermined circumstances. Therefore, it is plausible that data may exhibit high precision without necessarily possessing correctness or accuracy. The terminologies above are ubiquitous in the literature about the exposition of conventional analysis and experimentation techniques and have occasionally been employed (erroneously) in a substitutable manner (Mason & Gandhi, 1983). The standard deviation or variance of the results often quantifies the inverse relationship between precision and imprecision. As per the definition, precision excludes any systematic error or bias.

The measure of accuracy is frequently denoted by the inverse of the standard deviation or variance and encompasses any potential systematic error or bias. The concept of accuracy encompasses the stochastic imprecision arising from random error and any potential systematic error. A systematic error tends to increase the standard deviation (Zhang et al., 2020). In commercial coal quality measurement, systematic error holds greater significance than the accuracy expressed above. Once the magnitude of systematic error is minimized to the degree deemed insignificant, accuracy, and precision can serve as significant parameters in establishing a comprehensive sampling approach and analyzing outcomes obtained from diverse testing techniques. Typically, coal analysis does not involve estimating the bounds of precision, which refers to the degree of deviation from an authentic or hypothetical value. In contrast, precision is ascertained through collaborative testing initiatives. Determining precision in laboratory testing involves two key aspects: repeatability and reproducibility (Li et al., 2019). Repeatability pertains to the precision of test results when conducted in the same laboratory, typically by the same analyst, utilizing the same equipment and adhering to the prescribed methodology. On the other hand, reproducibility refers to the precision of test results when conducted in different laboratories. The values cited in testing methodologies represent the discrepancies between two outcomes that should be surpassed in a mere 5% of result pairs, equivalent to twice the standard deviation of many outcomes.

The absence of a statistical confidence level in the description of repeatability and reproducibility periods undermines the accuracy and precision of the specifications, thereby raising doubts about the extent of the underlying variance. In the scenario where the repeatability interval must not be surpassed, it would be necessary for the variance to attain a value of zero. From a pragmatic perspective, this task is arduous, if not unattainable. The variances, also known as standard deviations, hold significant importance in the execution of testing and sampling operations as they can be divided into distinct components linked to identifiable reasons for variation (Body & Chadwick, 2001). This enables the evaluation of the comparative significance of particular particulars regarding precision and accuracy. Regarding reproducibility, a factor of stochastic variability impacts the level of concurrence among laboratories, while it does not influence the level of concurrence within laboratories (repeatability).

The sampling variance constitutes a significant portion of the overall accuracy of results, being the most substantial individual component. This is a significant matter often overlooked by those lacking expertise in the subject. Various testing techniques outline the process for obtaining a comprehensive coal sample and the approach for determining the total variance for fixed weight increments of a specific coal type (Peng et al., 2012). The level of precision is such that in the event of multiple gross samples being taken from a lot of the shipment and an ash determination being conducted on the analysis sample of each gross sample, 95% of these determinations will be within a 10% range of the overall average of all determinations. Nevertheless, it is possible that certain circumstances may hinder the attainment of such accuracy, and it is imperative to maintain a proper perspective regarding the statement's efficacy.

Currently, in cases where multi-seam blended coal specimens contain mineral matter ranging from 10% to 30% by weight, a high degree of precision is necessary to determine the amount of ash remaining after combustion accurately. Failure to achieve such precision might result in a difference of 4 to 5% in the ash produced. To address these concerns, it is necessary to devise a sampling program that accounts for the possibility of variations in the analytical data. An appropriate program should entail the collection of specimens from multiple predetermined locations within the coal heap to account for variations in coal properties and mineral matter segregation during the coal's lifespan (Andres & Bona, 2005). The sampling properties of coal play a crucial role in using text-based techniques for generating sales data.

4.3. BIAS

The inclusion of the term "bias" in coal analysis serves the objective of ensuring the accuracy of the analytical data (Zhang et al., 2021). Comprehending the concepts of precision and bias in quantitative test methodologies is crucial to guarantee the precision of the data generated by analytical test methodologies.

Assessing bias in a coal collection procedure is crucial to coal analysis and holds considerable significance. The term "bias" denotes the presence of a consistent error or errors that hold practical significance.

Determining systematic error is conducted by computing the discrepancies among duplicate outcomes. To identify a systematic error, from a statistical perspective, it is imperative to decrease the precision boundaries of the mean to a value that is lower than a certain numerous of the standard deviation of the differences. For systematic error to be considered biased, it must attain a magnitude that holds practical significance (Zeng et al., 2017). The inadequate experimental design may result in a systematic error that could be practically significant due to multiple sources of error. The errors of omission observed in the data have the potential to cause confusion or mislead the reader, thereby highlighting the lack of reliability of the evaluation method(s).

The objective is not to eliminate all forms of bias but to mitigate bias to a tolerable degree that does not surpass a predetermined threshold in each instance. The test for bias may be formulated to verify the existence of bias if the likelihood of a bias of comparable magnitude is present (Siregar et al., 2017). Undoubtedly, the inherent characteristics of the problem are such that the verification of impartiality is unattainable.

The matters about relative bias or absolute bias necessitate careful consideration. Relative bias probably pertains to the evaluation of gross sample outcomes. In contrast, absolute bias is contingent upon the comparison with reference values that are devoid of bias and typically necessitates comparisons on an incremental basis (Yu et al., 2017).

The assessment of bias involves a series of fundamental procedures, which encompass the subsequent crucial stages:

i. A preliminary examination is carried out before conducting the test.

ii. The selection of the appropriate test method specifications is made.

iii. The implementation of comprehensive protocols for the execution of the testing methodology.

iv. This step involves collecting, processing, and analyzing preliminary test method increments (samples) (Zeng et al., 2017).

v. The determination of the necessary number of observations.

vi. The ultimate collection, processing, and analysis of the increment.

vii. The statistical evaluation and interpretation of the data are also crucial to the research process (Hao et al., 2016).

It involves the application of statistical methods to analyze and draw conclusions from data collected during the study. This step is essential in ensuring that the findings are reliable and valid (Zhao et al., 2017).

For each variable about coal constituents or properties that needs to be analyzed, it is necessary to allocate a corresponding test method. From a pragmatic standpoint, every element or characteristic is ascertained through a testing procedure that is frequently self-contained. In addition, it is recommended to assess all components of the coal on a dry basis, with the exclusion of any moisture, utilizing a standardized coal size. The size distribution of coal constituents is prone to errors due to size selectivity (Xue et al., 2009). Using screen tests to acquire size distribution data, particularly in the extremities of the size distribution (as per ISO, 1953), may occasionally be advantageous. However, size may not always be a suitable variable for testing.

4.4. REPORTING COAL ANALYSIS

Various methods can be employed to present analyses based on the humidity and ash content levels. The term "as-determined" denotes the moisture state of the specimen at the time of laboratory analysis (Potgieter-Vermaak et al., 2011). It is a common procedure to allow the sample to undergo air-drying, which results in the moisture content reaching a state of almost equilibrium with the lab's atmosphere. This is done to reduce the likelihood of experiencing gain or loss during sampling. The calculation of the as-received basis, which refers to the moisture condition of a sample upon arrival at the laboratory, is facilitated by the measurement of weight loss during air drying. Assuming that there is no alteration in moisture content during transportation from the sampling location to the laboratory, the statement above is equivalent to the as-sampled basis. Efforts to preserve the moisture content at the initial sampling level involve transportation via hermetically sealed vessels with enclosed plastic padding or in airtight plastic pouches.

Calculations of analyses on a dry basis depend on the assumption that the sample is devoid of any moisture content. The moisture content converts the ascertained data to a dry basis (Mo et al., 2014). Analytical data on a dry, ash-free basis is calculated based on the presumption that the sample contains no moisture or mineral matter. The values acquired through the process of moisture identification as well as ash determination, are utilized for conversion. Ultimately, the data computed on an equilibrium moisture basis is predicated on the moisture level that has been ascertained as equilibrium moisture.

Did You Know?

The concept of proximate analysis was first introduced by Francis H. Cron in 1901.

The presence of hydrogen and oxygen in moist substances may be subject to including or excluding the oxygen and hydrogen in the accompanying moisture. The analytical report must specify this aspect due to the variability in conversion factors. The factors above apply to both calorific values and the proximate analysis presented in Table 4.1 and the ultimate analysis outlined in Table 4.2 (Lu et al., 2021).

Table 4.1. Data Derived from Proximate Analysis

	Moisture	**Ash**	**Volatile**	**Fixed Carbon**
Air-Dried	8.23	4.46	40.05	47.26
Dry	–	4.86	43.64	51.50
As-Received	23.24	3.73	33.50	39.53

Source: Winefordner, Creative Commons License.

Table 4.2. Data Derived from Ultimate Analysis

Basis	**Carbon**	**Hydrogen**	**Nitrogen**	**Sulfur**	**Ash**	**Oxygen**[a]	**Moisture**	**Total (%)**
As-determined[b,c]	60.08	5.44	0.88	0.73	7.86	25.01	9.00	100.0
Dry	66.02	4.87	0.97	0.80	8.64	18.70	0.00	100.0
As-received[d]	46.86	6.70	0.69	0.57	6.13	39.05	(29.02)	100.0
As-received[e]	46.86	3.46	0.69	0.57	6.13	13.27	29.02	100.0

[a]By difference.

[b]After air-dry loss (22.00%) in accordance with ASTM D-2013.

[c]Hydrogen and oxygen include hydrogen and oxygen in sample moisture,

Mad.

[d]Hydrogen and oxygen include hydrogen and oxygen in sample moisture, Mar.

[e]Hydrogen and oxygen do not include hydrogen and oxygen in sample moisture, Mar.

Source: Winefordner, Creative Commons License.

The calculation of dry basis values can be determined by utilizing specific formulas when the percentages of hydrogen and oxygen include the moisture's hydrogen and oxygen content:

$$H_d = (H^1 - 0.1111M^1) \times \frac{100}{100 - M^1}$$

$$O_d = (O^1 - 0.8881M^1) \times \frac{100}{100 - M^1}$$

The weight percentage of oxygen and hydrogen on a dry basis denoted as Od and Hd, respectively, can be expressed in terms of the weight percentages of oxygen and hydrogen, or O_1 and H_1, respectively, for a given or identified weight percentage of moisture, M_1 (Kumar et al., 2020). The equations can be rearranged to derive solutions for H_1 and O_1, which enable calculating hydrogen or oxygen contents H_1 or O_1 at any given moisture level M_1.

Remember

The calorific value of coal can vary widely depending on its composition.

During the thermal conversion of coal to ash, the mineral matter (Ode, 1963) undergoes a weight reduction. This is attributed to the dehydration of clays, the decarboxylation of carbonate minerals like calcite, and the oxidation of pyrite (FeS_2) to ferric oxide (Fe_2O_3) (Hatton & Fardell, 2012). Furthermore, the coal undergoes a conversion process that transforms any chlorine present into hydrogen chloride. However, it is worth noting that the alteration in mass may not be substantial.

The Parr formulas (ASTM D-388) are utilized to determine analyses and calorific values on a mineral-matter-free basis while accounting for pyrite and additional mineral matter through appropriate corrections. The total sulfur content of coal is often used as a proxy for the quantity of pyrite present, despite the possibility of inaccuracies. This approach has been observed to exhibit strong correlations in mineral matter investigations. It is assumed that the residual mineral content is equivalent to 1.08 times the mass of the ash that is free from iron oxide (Hunting & Weeks, 1993):

$$mm = 1.08A + 0.55S$$

The weight percentages of mineral content, ash, and total sulfur are represented by mm, A, and S, respectively.

The information above is essential for the computation of variables involved in categorizing coal based on its quality, including the mineral-free volatile matter in a dry state and the mineral-free gross calorific value (GCV) in a moist state. To obtain accurate measurements of volatile matter and fixed carbon, assuming that 50% of the sulfur by weight is volatilized during the volatile matter test is imperative. This is necessary to ensure that the sulfur does not contaminate the volatile organic matter, nor by the loss from clays and minerals made of carbonate (Bacon & Nash, 2012):

$$FC_{dmmf} = \frac{100(FC - 0.15S)}{100 - (M + 1.08A + 0.55S)}$$

$$VM_{dmmf} = 100 - FC$$

The variables FC_{dmmf} and VM_{dmmf} pertain to fixed carbon and volatile matter on a dry, mineral-matter-free basis. Meanwhile, the variables FC, M, A, and S denote the calculated fixed carbon, water content, ash, and total sulfur.

The Parr formula utilized for determining the calorific value of moist, mineral-matter-free coal incorporates the naturally occurring moisture that exists in the coal seam, also known as natural bed moisture or capacity moisture, as the basis for moisture content (Harris et al., 2014):

$$\text{moist, mineral-matter-free Btu} = \frac{100(Btu - 50S)}{100 - (1.08A + 0.55S)}$$

where; Btu (Btu/lb), A (ash-to-weight ratio), and S (sulfur-to-weight ratio) are, they're all using natural beds that are damp (Figure 4.2).

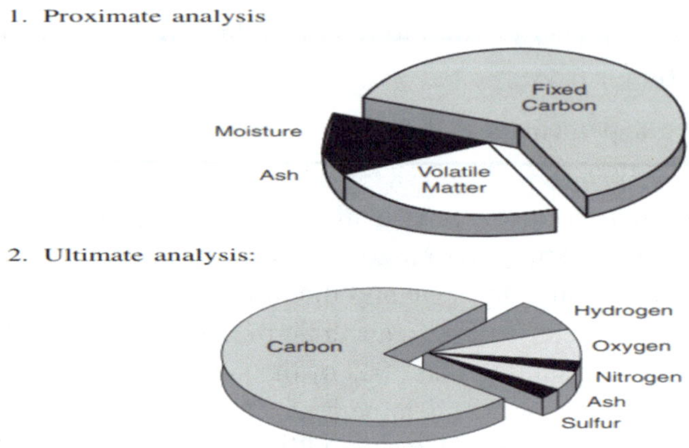

Figure 4.2. *Data types were obtained from (a) proximate analysis; and (b) ultimate analysis.*

Source: James G. Speight, Creative Commons License.

4.5. INTERRELATIONSHIPS OF THE DATA

Similar correlations exist between the various qualities of coal and factors like the depth at which a reservoir is buried, just as there is one between petroleum's many properties and various parameters (Deniz & Umucu, 2013). Variations in the proportions of hydrogen or oxygen to carbon and between each other were additionally identified. It should be emphasized, nonetheless, that many of the available publications indicate the variance of analytical findings or test results using elemental carbon content rather than rank, which can only be roughly equated with it.

Other correlations exist, such as differences in the volatile matter in vitrinite macerals found at various depths and variations in the natural bed dampness with burial depth. The latter finding (i.e., a reduction in volatile matter with the seam's burial depths) contrasts sharply with similar findings for petroleum, which show a rise in lower-molecular-weight content as reservoir depth increases (Newman, 1985).

The production of hydrogen-rich species (such as the components of the gasoline fraction) in deeper petroleum reservoirs appears to be in stark contrast to the propensity toward a carbon-rich material in deeper coal seams (Neavel et al., 1986). The characteristics of the raw material and the various maturation procedures significantly influence the end product's nature. Last, but not least, it is also feasible to show how the results from the proximate evaluation and the calorie content relate to coal rank.

4.6. THE FUTURE

Conventionally, wet analysis or standard bench-scale laboratory equipment have been used in coal analysis. This pattern is still present and could last another 10 or 20 years. However, the creation of a new generation of coal analysis devices and the required calibration of such instruments has occurred recently due to the development of microprocessors and microcomputers. In particular, automatic instrumentation has been developed to quickly determine the temperatures of ash fusion, water content, ash, volatile matter, nitrogen, carbon, hydrogen, and sulfur compared to most traditional laboratory bench processes (Kavalov & Peteves, 2007).

Several such devices have been created to measure the three elements in diverse samples simultaneously. Of course, the basic needs for the equipment are that they enable the quantitative measurement of these gases in an adequate gas stream as well as the entire transformation of the nitrogen, carbon, and hydrogen in coal to water vapor, carbon dioxide, and elemental nitrogen.

The tiny sample size utilized in the study is a drawback of various instrumental procedures for identifying carbon, hydrogen, and nitrogen (Yuan, 2018). Common sample sizes for a few of the equipment may range from 1 to 3 mg on the best days. However, the system's accuracy may be called into doubt. As long as effluents don't flood the system or overwhelm it to the point that the detecting equipment cannot function, other systems that employ 100-mg samples might be selected. The likelihood that the sample represents the amount of coal being studied does, however, rise with a bigger sample size.

The majority of techniques utilized by the new analytical all-in-one instruments are empirical, and the applicability and quality of the standards used to standardize the instruments significantly impact the results' correctness (Zhang et al., 2017).

4.7. SUMMARY

Coal analysis plays a significant role in coal mining, processing, utilization, and environmental management. It involves various techniques and tests to assess the quality, suitability, and potential uses of coal. The key aspects of coal analysis include proximate analysis, ultimate analysis, calorific value, ash analysis, sulfur content and trace element analysis. The information obtained from coal analysis guides decisions related to coal quality control, blending, pricing, and determining the appropriate coal for specific applications.

REFERENCES

1. Andres, J. M., & Bona, M. T., (2005). Analysis of coal by diffuse reflectance near-infrared spectroscopy. *Analytica Chimica Acta, 535*(1–2), 123–132.
2. Bacon, W., & Nash, C., (2012). Playing the media game: The relative (in) visibility of coal industry interests in media reporting of coal as a climate change issue in Australia. *Journalism Studies, 13*(2), 243–258.
3. Balachandran, M., (2014). Role of infrared spectroscopy in coal analysis – An investigation. *American Journal of Analytical Chemistry, 3*(1), 3–7.
4. Body, D., & Chadwick, B. L., (2001). Simultaneous elemental analysis system using laser induced breakdown spectroscopy. *Review of Scientific Instruments, 72*(3), 1625–1629.
5. Bona, M. T., & Andrés, J. M., (2007). Coal analysis by diffuse reflectance near-infrared spectroscopy: Hierarchical cluster and linear discriminant analysis. *Talanta, 72*(4), 1423–1431.
6. Deniz, V., & Umucu, Y., (2013). Interrelationships between the Bond grindability with physicomechanical and chemical properties of coals. *Energy Sources, Part A: Recovery, Utilization, and Environmental Effects, 35*(2), 144–151.
7. Donahue, C. J., & Rais, E. A., (2009). Proximate analysis of coal. *Journal of Chemical Education, 86*(2), 222–224.
8. Dooher, J. P., (2005). Fundamental considerations for coal slurry atomization. *Atomization and Sprays, 15*(5), 2–6.
9. Eskanazy, G., Finkelman, R. B., & Chattarjee, S., (2010). Some considerations concerning the use of correlation coefficients and cluster analysis in interpreting coal geochemistry data. *International Journal of Coal Geology, 83*(4), 491–493.
10. Han, K. W., Dixit, V. B., & Wen, C. Y., (1978). Analysis and scale-up consideration of bituminous coal liquefaction rate processes. *Industrial and Engineering Chemistry Process Design and Development, 17*(1), 16–21.
11. Hao, Y., Liu, Y., Weng, J. H., & Gao, Y., (2016). Does the environmental Kuznets curve for coal consumption in China exist? New evidence from spatial econometric analysis. *Energy, 114*(1), 1214–1223.
12. Harris, J., Kirsch, P., Shi, M., Li, J., Gagrani, A., Krishna, E. S. A., & Cliff, D., (2014). *Comparative Analysis of Coal Fatalities in Australia, South Africa, India, China, and USA, 2006–2010* (Vol. 1, pp. 3–6).
13. Hatton, W., & Fardell, A., (2012). New discoveries of coal in Mozambique – Development of the coal resource estimation methodology for

International Resource Reporting Standards. *International Journal of Coal Geology, 89*(1), 2–12.
14. Hou, Z., Wang, Z., Yuan, T., Liu, J., Li, Z., & Ni, W., (2016). A hybrid quantification model and its application for coal analysis using laser induced breakdown spectroscopy. *Journal of Analytical Atomic Spectrometry, 31*(3), 722–736.
15. Hunting, K. L., & Weeks, J. L., (1993). Transport injuries in small coal mines: An exploratory analysis. *American Journal of Industrial Medicine, 23*(3), 391–406.
16. Kavalov, B., & Peteves, S. D., (2007). *The Future of Coal* (Vol. 1, pp. 2–8). Institute for Energy of European Commission.
17. Kiss, L. T., & King, T. N., (1977). Expression of results of coal analysis: the case for brown coals. *Fuel (United Kingdom), 56*(3), 5–10.
18. Kumar, P., Gupta, S., & Gunda, Y. R., (2020). Estimation of human error rate in underground coal mines through retrospective analysis of mining accident reports and some error reduction strategies. *Safety Science, 123*(1), 104–555.
19. Li, W., Dong, M., Lu, S., Li, S., Wei, L., Huang, J., & Lu, J., (2019). Improved measurement of the calorific value of pulverized coal particle flow by laser-induced breakdown spectroscopy (LIBS). *Analytical Methods, 11*(35), 4471–4480.
20. Lu, C., Dasgupta, P., Cameron, J., Fritschi, L., & Baade, P., (2021). A systematic review and meta-analysis on international studies of prevalence, mortality, and survival due to coal mine dust lung disease. *Plos One, 16*(8), 255–617.
21. Mason, D. M., & Gandhi, K. N., (1983). Formulas for calculating the calorific value of coal and coal chars: Development, tests, and uses. *Fuel Processing Technology, 7*(1), 11–22.
22. Mo, J., Wang, L., Au, W., & Su, M., (2014). Prevalence of coal workers' pneumoconiosis in China: A systematic analysis of 2001–2011 studies. *International Journal of Hygiene and Environmental Health, 217*(1), 46–51.
23. Neavel, R. C., Smith, S. E., Hippo, E. J., & Miller, R. N., (1986). Interrelationships between coal compositional parameters. *Fuel, 65*(3), 312–320.
24. Newman, J., (1985). *Paleoenvironments, Coal Properties and Their Interrelationship in Paparoa and Selected Brunner Coal Measures on the West Coast of the South Island* (Vol. 1, pp. 4–7).

25. Peng, B., Wu, D., Lai, J., Xiao, H., & Li, P., (2012). Simultaneous determination of halogens (F, Cl, Br, and I) in coal using pyrohydrolysis combined with ion chromatography. *Fuel, 94*(1), 629–631.
26. Potgieter-Vermaak, S., Maledi, N., Wagner, N., Van Heerden, J. H. P., Van Grieken, R., & Potgieter, J. H., (2011). Raman spectroscopy for the analysis of coal: A review. *Journal of Raman Spectroscopy, 42*(2), 123–129.
27. Sheta, S., Afgan, M. S., Hou, Z., Yao, S. C., Zhang, L., Li, Z., & Wang, Z., (2019). Coal analysis by laser-induced breakdown spectroscopy: A tutorial review. *Journal of Analytical Atomic Spectrometry, 34*(6), 1047–1082.
28. Siregar, I., Niu, Y., Mostaghimi, P., & Armstrong, R. T., (2017). Coal ash content estimation using fuzzy curves and ensemble neural networks for well log analysis. *International Journal of Coal Geology, 181*(1), 11–22.
29. Ward, C. R., (2002). Analysis and significance of mineral matter in coal seams. *International Journal of Coal Geology, 50*(1–4), 135–168.
30. Ward, C. R., (2016). Analysis, origin, and significance of mineral matter in coal: An updated review. *International Journal of Coal Geology, 165*(1), 1–27.
31. Warwick, P. D., (2005). Coal systems analysis: A new approach to the understanding of coal formation, coal quality and environmental considerations, and coal as a source rock for hydrocarbons. *Geological Society of America Special Papers, 387*(1), 1–8.
32. Xue, S., Liu, T., Zhou, Q., Xu, T., & Hu, H., (2009). Experimental investigation on NOx emission and carbon burnout from a radially biased pulverized coal whirl burner. *Fuel Processing Technology, 90*(9), 1142–1147.
33. Yu, H., Chen, H., & Long, R., (2017). Mental fatigue, cognitive bias, and safety paradox in Chinese coal mines. *Resources Policy, 52*(1), 165–172.
34. Yuan, J., (2018). The future of coal in China. *Resources, Conservation, and Recycling, 129*(1), 290–292.
35. Zeng, G., Sun, S., Yang, X., Zhao, Y., Zhao, Z., Ye, Z., & Gao, J., (2017). Effect of the primary air velocity on ignition characteristics of bias pulverized coal jets. *Energy and Fuels, 31*(3), 3182–3195.
36. Zeng, G., Sun, S., Zhang, W., Zhao, Z., & Zhao, Y., (2017). Effects of bias concentration ratio on ignition characteristics of parallel bias pulverized coal jets. *Energy and Fuels, 31*(12), 14219–14227.
37. Zhang, L., Gao, W., Chiu, Y. H., Pang, Q., Shi, Z., & Guo, Z., (2021). Environmental performance indicators of China's coal mining industry: A

bootstrapping Malmquist index analysis. *Resources Policy, 71*(1), 101991.

38. Zhang, X., Winchester, N., & Zhang, X., (2017). The future of coal in China. *Energy Policy, 110*(1), 644–652.
39. Zhang, Y., Dong, M., Cheng, L., Wei, L., Cai, J., & Lu, J., (2020). Improved measurement in quantitative analysis of coal properties using laser induced breakdown spectroscopy. *Journal of Analytical Atomic Spectrometry, 35*(4), 810–818.
40. Zhao, J., Zhou, L., Zhou, W., Ren, H., Yu, Y., Wang, F., & Ma, T., (2021). Techno-economic analysis and comparison of coal-based chemical technologies with consideration of water resources scarcity. *Energy Strategy Reviews, 38*(1), 100754.
41. Zhao, Y., Zeng, G., Zhang, L., Zhang, Y., Sun, S., & Wei, C., (2017). Effects of fuel properties on ignition characteristics of parallel-bias pulverized-coal jets. *Energy and Fuels, 31*(11), 12804–12814.
42. Zhu, Q., (2014). *Coal Sampling and Analysis Standards* (Vol. 143, No. 1, pp. 4–7). IEA Clean Coal Center, London, United Kingdom.
43. Żogała, A., (2014). Critical analysis of underground coal gasification models. Part I: Equilibrium models – literary studies. *Journal of Sustainable Mining, 13*(1), 22–28.

CHAPTER

5 Physical and Electrical Properties of Coal

Contents

INTRODUCTION

Many physical and electrical properties can provide information regarding coal's application (Van Krevelen, 1957; Evans & Allardice, 1978). These properties can be determined in the same way coal evaluation can be accomplished by identifying numerous prominent features. In point of fact, there are certain features of organic substances which provide beneficial information regarding the behavior of organic materials in the environment (Lyman et al., 1990). This is yet another argument in favor of studying coal's characteristics.

In order to properly characterize and understand coal and its behavior, one must also have a solid know-how of the coal's electrical characteristics. Electrical qualities are helpful in a variety of operations, including maintenance and carbonization. They are also of particular relevance in electro-linking, which is used to promote permeability. It also identifies areas with distinct physical characteristics during *in situ* coal gasification. Both of the applications are of exceptional interest. These aspects of coal's makeup and behavior play a significant role in determining how it should be processed and used. Thus, understanding these aspects is essential.

The physical properties of coal can be better understood by considering it as a fragmented medium composed of a graphite-like substance embedded in an insulating organic matrix (Speight, 1994 and related sources). This proposal suggests that the dispersed structure of high-rank coals is of particular importance in this regard. This concept has been proposed in a broad and overarching manner. Indeed, there have been indications of coal's resemblance to graphite, primarily through X-ray diffraction analyses (Speight, 1978, and sources cited therein). It is plausible that this approach could be utilized to elucidate the properties of coal. Nonetheless, in the event that this is the case, it is imperative to fully characterize the precise function that the aromatic systems with reduced scope serve. Notwithstanding this fact, it is widely held that it introduces novel avenues of inquiry pertaining to the conduct of coal.

At first glance, there might not appear to be much of a connection, if any at all, between the way coal behaves physically and the way it reacts chemically, but in reality, the opposite is very much the case. According to Walker (1981), one of the most important factors when assessing the chemical reactivity of coal is the coal's pore size, which is technically considered a physical feature. Coal should be treated differently before or during conversion processes depending on the chemical reactions that cause the expansion and caking of coal(s). This is because chemical effects cause coal(s) to become more viscous. This is the rationale behind investigating and assessing these extremely significant qualities of coal.

When trying to identify the coal's physical characteristics, there is frequently a substantial amount of controversy regarding the validity of the data produced from measurements carried out using two or more distinct analytical approaches. The most important questions to ask are whether or not the specimen was properly prepared and measured, as well as whether or not the analytical parameters were applied correctly in the step(s) of data handling, given that the procedures and machinery that was utilized was able to provide finest data. Therefore, the idea that different methods can produce different answers that are still true can be applied to the computation of substantial qualities, for instance, particle size, density, porosity, and surface area. These are all examples of properties that can be measured.

5.1. DENSITY AND SPECIFIC GRAVITY

There are five different ways to estimate density for porous substances like coal: apparent density, real density, bulk density, particle density, and in-place density. The mass of the coal is divided by the volume it occupies to determine its true density. However, calculating coal's mass may be thought of as being quite simple, but calculating coal's volume provides some challenges. Volume cannot be uniformly defined in a single sentence. In fact, the experimental procedure used to quantify volume and density must apply to measure principles that adhere to the chosen criteria.

In order to obtain an accurate measurement of the true density, it is necessary to completely saturate the pore structure with a liquid that does not come into proximity to the solid material. None of the fluids available in the market completely meet these criteria. Although helium has traditionally been considered the optimal option due to its negligible retention by coal at ambient temperature, research conducted by Berkowitz (1979) and other cited sources suggests that a portion of the pore structure may be inaccessible to helium. Consequently, using helium to measure coal density may lead to a deviation from the true density.

Therefore, it is not shocking that coal density varies and is type-dependent. For instance, bituminous coal has a density of around 1.35, lignite has a density of about 1.25, and anthracite has a density of about 1.55. The density of coal is normally calculated by the fluid displacement, but owing to the permeability of coal, the density values observed vary depending on the fluids employed, and various fluids may be needed to crush different types of coal (Agrawal, 1959; Mahajan & Walker, 1978). As a result, such generalizations should be treated with caution.

By employing meticulous methodology, a standard testing protocol designed to ascertain the actual density of coal may also be utilized to calculate the genuine density of coal. The proposed methodology involves the estimation of permeability through the utilization of specific gravity information to ascertain the apparent and true specific gravities of bundled coal exceeding 25 mm in size. The apparent true and apparent specific gravity of coal, as determined by this testing methodology, is influenced by various factors, including the type of coal subjected to carbonization, as well as the operational and preparatory aspects of the process, such as charge bulk density, heat rate, and crushing level.

The trial method calls for employing a 13-mm square mesh wire cloth cage. Because collecting a representative sample is challenging, one should be careful while picking pieces representing the full visible region. Furthermore, during sample preparation, it may be required to air-dry the specimen before crushing or grinding it. Whatever the situation is, one should be extremely careful while picking the size-reduction machinery that will not erode or intrude undesired contaminants into the specimen.

The phrase "helium density" is frequently employed to denote the actual density, which is commonly ascertained via helium displacement. The utilization of helium is attributed to its capacity to infiltrate all the pores present in a particular specimen without causing any chemical response. The direct-pressure technique entails the insertion of a predetermined quantity of helium and a quantified coal sample into the specified volume apparatus. The determination of coal volume can be facilitated by the helium pressure at a particular temperature. The indirect method employs mercury as a compensatory measure for the reduction in helium levels caused by the introduction of coal.

The real density (which is actually the helium density) of coal has been linked to its component composition. Hence,

$$\text{density}_{He} = 1.534 - 0.05196H + 0.007375O - 0.02472N + 0.003853S$$

where; density_{He} is the helium density (g/cm^3) and H, O, N, and S are the respective weight percentages.

The density of coal is a significant parameter, not only for elucidating the molecular architecture of coal but also due to the correlation between density and porosity, which underscores the importance of prioritizing density determination. The measurement of porosity, in turn, gives significant understanding regarding the technical conduct of coal as it approaches its eventual use. Individual coal particle porosity is calculated using particle density.

A gas pycnometer or mercury porosimetry can be used to determine the density of coal particles. It is important to recognize, however, that the density values acquired with different gases might vary widely; for instance, nitrogen may acquire greater density values than those acquired with helium. Variations may be ascribed to nitrogen adsorption on the coal surface, influencing density measurement.

A gas pycnometer detects pressure changes caused by the movement of gas by a solid body. A baseline is established by expanding a known amount of gas into a vacant chamber and monitoring the pressure. Subsequent to the placement of the specimen, the chamber is hermetically sealed. An equivalent quantity of gas is introduced into the chamber at identical pressure, and subsequently, the pressure is recalculated. The application of the gas law enables the computation of the volume of a given sample by utilizing the disparity in pressures and the established volume of the unoccupied chamber. The gas pycnometer has good accuracy and precision in estimating density, but the method is heavily reliant on the specimen's hygiene and the gas's pureness. Small open pores are extremely accessible. When the sample contains small, open pores, the volume calculated by this approach will be less than that determined using other liquid displacement methods.

Mercury is a nonwetting liquid, meaning it cannot enter a pore without being pushed in from the outside. Mercury has the ability to create a transient bridge over apertures, crevices, and fractures due to its elevated surface tension until an adequate external force is exerted to facilitate its infiltration. Under standard atmospheric pressure conditions, it can be observed that mercury exhibits a limited ability to infiltrate pores that possess a diameter of less than 6 micrometers. The existence of mercury leads to the creation of a liquid envelope that conforms closely to an object. The adhesion of mercury to the surface characteristics of an entity is contingent upon the level of force exerted on it. The sample demonstrates a reduction in the quantity and dimensions of voids it accommodates as the pressure is elevated. Mercury has been effectively injected into pores with a diameter as small as 0.003 m under a pressure of 60,000 psi. This substance typically occupies the vast majority of the pore volume of the majority of materials. The process of ascertaining the bulk density and skeleton density is commonly executed through the measurement of the volume of mercury that is displaced at the minimum and maximum pressures. The quantification of the grain volume in a powder can be accomplished through the deduction of the interparticle void volume from the bulk volume.

The mercury porosimeter is the most popular tool for measuring density according to pore size and hence estimating pore volume distribution. Therefore, pore volume is typically an afterthought when analyzing data collected using porosimetry.

The determination of coal's apparent density is achieved through the immersion of a coal sample of known weight in a liquid, followed by the measurement of the volume of liquid displaced, according to the pycnometer method. The liquid utilized must satisfy the specified criteria for this method to be effective. The four characteristics of penetrating liquid coal are as follows: (i) the act of moistening the outermost layer of the coal; (ii) a lack of significant integration with the coal surface; (iii) an absence of swelling; and (iv) the ability to permeate the pores of the coal.

Meeting these criteria is challenging, if not unattainable, as evidenced by the inconsistencies in empirical data obtained using chemicals such as carbon tetrachloride, water, methanol, benzene, and other fluids. Consequently, it is imperative to specify the liquid employed in determining density through the utilization of the pycnometer method. Moreover, the determination may necessitate a 24-hour time frame owing to the need for the liquid to infiltrate the pore network of the coal to its maximum extent.

Mahajan & Walker (1978) defined particle density as the mass of a unit volume of solid material, which includes pores and fissures. Particle density can be determined through the utilization of three distinct techniques, namely silanization (Ettinger & Zhupakhina, 1960), mercury displacement (Gan et al., 1972), and gas flow (Ergun, 1951).

The variability of coal density is contingent upon the proportion of carbon present, as illustrated in Figure 5.1. The observed phenomenon can be ascribed to the reduction in the volume of helium that is adsorbed within the pores of coal, in addition to the intermolecular interactions that occur between coal and methanol. Moreover, it has been observed that the density of methanol exceeds that of helium, resulting in a total volume that is significantly less than the sum of the individual volumes.

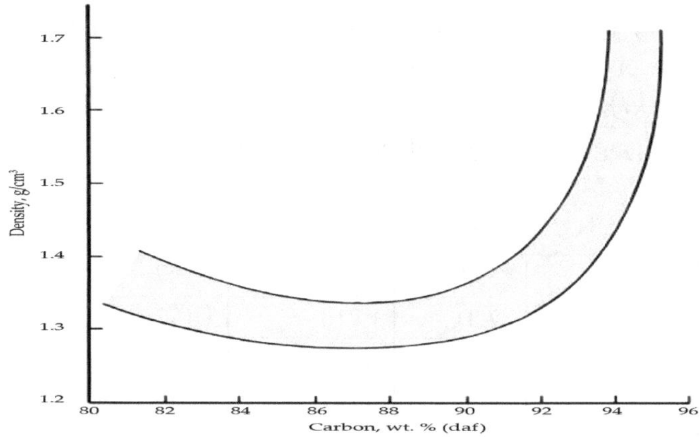

Figure 5.1. *Density varies with the amount of carbon.*

Source Berkowitz, Creative Commons License.

Although water has a lower density than helium, studies have demonstrated that coal containing over 85% w/w carbon exhibits greater hydrophobicity than lower-rank coals. In the case of coal containing 80 to 84% carbon, there's generally minimal or no discrepancy between the densities of helium and water. The hydrophobicity of coal is more closely associated with its moisture content than its carbon content. Additionally, the moisture/carbon molar ratio exhibits a stronger correlation with hydrophobicity than the hydrogen/carbon or oxygen/carbon atomic ratios. The strength of both of these correlations surpasses that of the relationship between hydrogen/carbon and oxygen/carbon. The available literature suggests a correlation between the hydrophobicity of coal and its moisture content (Labuschagne, 1987; Labuschagne et al., 1988).

An additional noteworthy phenomenon is an inclination for coal density to show a minimal threshold at approximately 85% weight-to-carbon ratio. As an illustration, coal samples containing 50–55% w/w carbon exhibit a density of approximately 1.5 g/cm^3. Conversely, coal specimens with an 85% carbon content display a lower density of 1.3 g/cm^3. However, coal with a 97% carbon content demonstrates a higher density of approximately 1.8 g/cm^3. Graphite's 2.25 g/cm^3 density exhibits a similar trend compared to other materials.

While minor discrepancies may exist, the typical sequence of maceral density for those with comparable carbon content is exinite, vitrinite, and micrinite. The density values of various coal macerals were determined and subsequently published in Table 5.1.

The density of a specific maceral may exhibit slight variations based on the carbon proportion, as depicted in Figure 5.2 (van Krevelen, 1957; Berkowitz, 1979).

Table 5.1. The Density of Macerals Regarding Alcohol and Helium

Macerals	% C	D_{CH3OH}	$d_{He(calc)}$
Exinites	85.49	1.201	1.187
	87.41	1.213	1.193
	89.10	1.288	1.267
	89.29	1.347	1.325
	83.5	1.345	1.304
	85.74	1.334	1.304
Vitrinites	88.36	1.317	1.295
	88.84	1.368	1.338
	86.77	1.463	1.435
	87.98	1.415	1.386
Micrinites	89.59	1.414	1.389
	89.78	1.413	1.385

Source: Kroger, Creative Commons License.

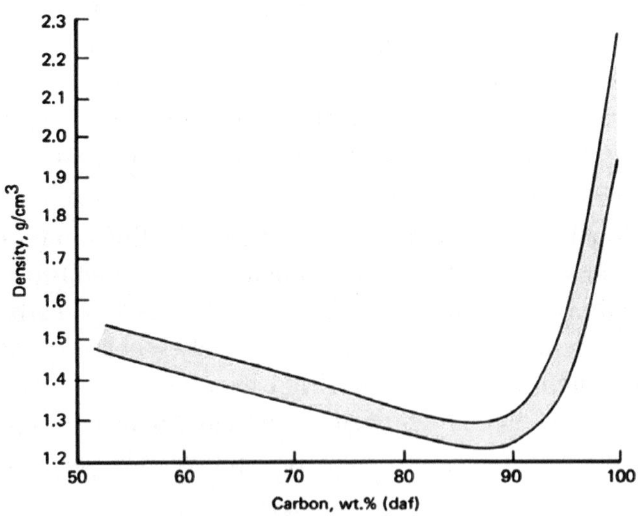

Figure 5.2. *The relationship between the amount of carbon in vitrinite and its density.*

Source Berkowitz, Creative Commons License.

The definition of bulk density pertains to the ratio of the mass of a collection of coal particles within a given container to the volume of said container. The role of various factors such as true density, particle size and shipping, particle shape, surface wetness, and degree of compaction is significant. The aforementioned criterion is commonly employed in the strategic development of infrastructure for the purposes of coal processing, transportation, and storage. The objective of this testing procedure is to evaluate the mass of coal that has been crushed and compacted, as well as the mass of coal that has not been compacted, for applications such as charging coke furnaces. The weight per cubic foot of coal is primarily influenced by two factors: the moisture content and the size distribution of the coal, in addition to its quality. In addition to the measurement of cubic foot weight, an accurate determination of cubic foot weight necessitates the inclusion of both moisture content measurement and sieve analysis of the coal. Simultaneous provision of this information is recommended. During the process of collecting the gross sample, it is recommended to store the increments of the specimen in a jar that is both watertight and equipped with a securely fitting lid. This measure will effectively mitigate the occurrence of moisture loss.

As demonstrated in Table 5.2, the utilization of a seam's *in-situ* density, commonly referred to as bank density, allows for the quantification of coal in terms of either ton per acre per foot of seam thickness or tons per square mile per foot of seam thickness. The determination of in-place density necessitates the utilization of water-saturated samples to account for the equilibrium moisture that arises in the in-place (or seam) condition (Berkowitz, 1979).

5.2. POROSITY AND SURFACE AREA

The porosity is the percentage of a coal's volume that is made up of pores (Berkowitz, 1979). Porosity can be calculated using equilibrium moisture content. Since coal is a porous substance, its behavior during mineral extraction, manufacturing, and consumption can be significantly influenced by permeability.

Although the pace at which methane can escape out of the coal is dictated by porosity, and there may be an impact during preparatory operations in regard to mineral matter elimination, the primary impact of the porous nature of coal is evident during coal consumption.

Table 5.2. Representation of In-Place Density of Coal

Rank	Tons per Acre per Foot of Thickness	Tons per Square Mile per Foot of Thickness
Anthracite	2,310	1,479 × 106
Semianthracite	2,039	1,305 × 106
Bituminous	1,903	1,218 × 106
Subbituminous	1,767	1,131 × 106
Lignite	1,631	1,044 × 106

Source: Baughman, Creative Commons License.

The determination of real specific gravity is used to calculate porosity, which is calculated from the relationship:

$$\text{porosity} = 100 - 100\left(\frac{\text{apparent specific gravity}}{\text{true specific gravity}}\right)$$

Another way to determine porosity is to measure the density of coal using helium displacement and mercury displacement. As a result, the porosity of coal is determined using the relationship:

$$P = 100 \times \text{density}_{Hg}\left(\frac{1}{\text{density}_{Hg}} - \frac{1}{\text{density}_{He}}\right)$$

where; P denotes porosity; Hg denotes mercury density; and He denotes helium density.

Remember

The hardness of coal can be assessed using the Mohs scale of mineral hardness.

The pore size (pore volume) distribution can be calculated by measuring the apparent density of coal in liquids of varying but known dimensions. The relationship can be used to compute the open pore volume (V) or the pore volume available to a certain fluid.

$$V = \frac{1}{\text{density}_{Hg}} - \frac{1}{\text{density}_{a}}$$

where; density_{Hg} is the mercury density, and density is the apparent density in the fluid under consideration.

The given relationship can be utilized to determine the pore volume.

$$V_p = \frac{1}{\text{density}_{Hg}} - \frac{1}{\text{density}_{He}}$$

where; density_{Hg} is the mercury density; and density_{He} is the helium density.

Coal's pore structures consist of micropores that have a maximum size of approximately 100 A and macropores that are larger than 300 A (Gan et al., 1972; Mahajan & Walker, 1978). The theory regarding the porosity of coals, which suggests that many pores have measurements that exceed 100-angstrom units in diameter but restricted access due to tiny holes that remove nitrogen at low temperatures, has been challenged by Kalliat et al.'s (1981) research on the same topic using small-angle X-rays. The authors have expressed doubts about the aforementioned theory, citing that the data do not provide sufficient evidence to support it. According to prevailing beliefs, the high surface area values obtained from adsorption tests can be attributed to the presence of numerous pores, wherein the minimum pore size is not more than 30 A.

Furthermore, there are indications that diminutive compounds such as methanol adhere to coal through a process that is specific to the site (Ramesh et al., 1992). In such circumstances, it is observed that the adsorption process initiates at locations with elevated energy levels. Nevertheless, as the adsorption proceeds, the adsorbate tends to adhere to the surface rather than other methanol molecules. Empirical evidence supports the occurrence of both physical and chemical adhesion phenomena in these contexts. A barrier to the adsorption process appears to exist when the coverage is below a monolayer. The potential influence of these findings on studies pertaining to surface area and distribution of pores is yet to be determined. The activation obstacle, a phenomenon of great interest, could potentially impact the interpretation of surface effects observed during coal combustion (Speight, 1994). In addition, it has been posited that investigations into the adsorption of minute molecules on coal have substantiated its copolymeric configuration (Milewska-Duda, 1991).

It has been observed that the permeability of coal diminishes with an increase in its carbon content (as depicted in Figure 5.3) (King & Wilkins, 1944; Berkowitz, 1979). The lowest permeability is observed in coals with approximately 89% w/w carbon. Subsequent to this, there is a discernible increase in permeability, which has been previously noted in connection with coal density (*refer* Figure 5.1). There appears to be a correlation between the carbon content (rank) of coal and the porosity type. As an illustration, coals with lower carbon ranks are generally characterized by a higher frequency of macropores. In contrast, coals with higher carbon ranks tend to possess a greater quantity of micropores. As a result, it can be observed from Figure 5.4 that the pore volume of coal diminishes with an increase in carbon content. Furthermore, the coal's surface area demonstrates fluctuations ranging from 10 to 200 m^2/g and displays analogous variations.

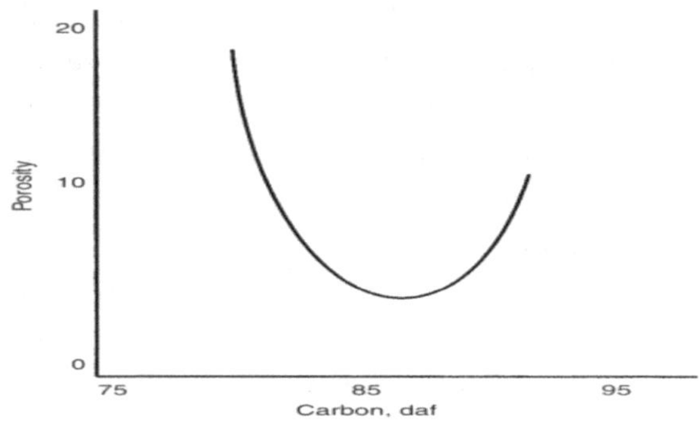

Figure 5.3. *Variation of porosity with carbon content.*

Source James G. Speight, Creative Commons License.

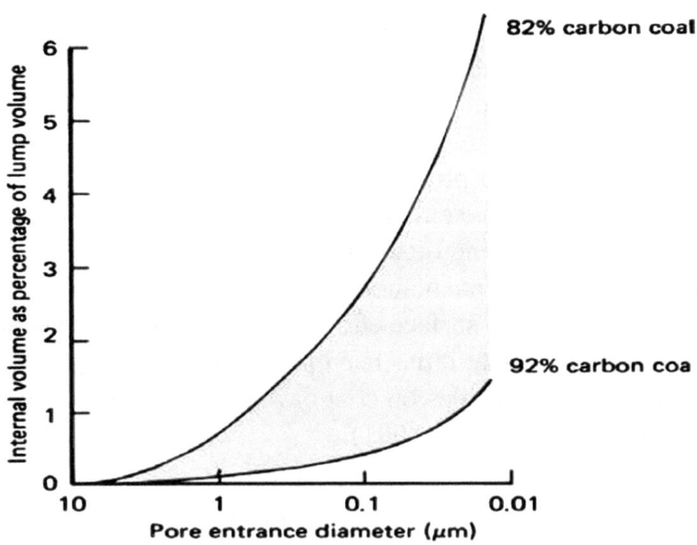

Figure 5.4. *Variation of pore distribution with carbon content.*

Source Dorman, Creative Commons License.

The determination of the pore size distribution within a coal specimen can be achieved through the utilization of mercury intrusion porosimetry, wherein the coal is subjected to increasing pressure while immersed in mercury. The phenomenon of surface tension imposes a restriction on the ability of mercury to infiltrate pores that possess a diameter less than a certain threshold value, denoted as d, under any given pressure, p.

$$p = \frac{4\sigma \cos\theta}{d}$$

where; σ is the surface tension; and θ the angle of contact.

The fluctuation of surface area with pore size can be elucidated through the measurement of mercury intrusion into coal under gradual pressure increases, as depicted in Figure 5.5 (van Krevelen, 1957; Berkowitz, 1979). Nonetheless, the pore volume calculated by this technique is notably inferior to that derived from helium density assessments, indicating that coal is composed of two discrete pore systems: (i) a set of macropores that can be penetrated by mercury under pressure; and (ii) a set of micropores that are not penetrable by mercury but can be entered by helium. The distribution of micropore diameters can be examined using liquids with different molecular sizes. Nevertheless, the specific function or purpose of these micropores within the structural model of coal remains unknown. However, there have been proposals suggesting that coal may operate similarly to a molecular sieve.

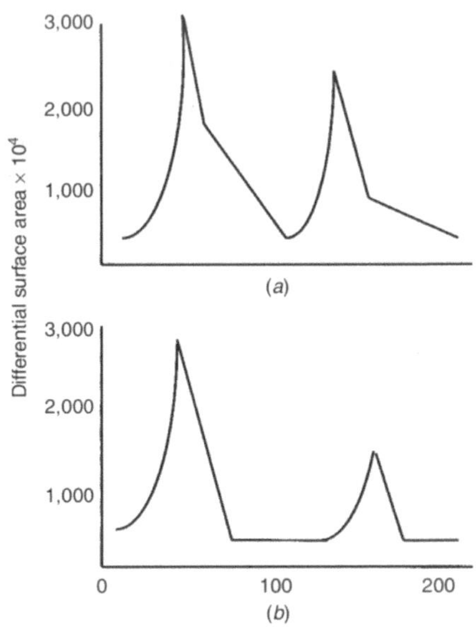

Figure 5.5. *The relationship between surface area and pore size variability.*

Source Berkowitz, Creative Commons License.

5.3. REFLECTANCE

Coal exhibits optical properties, despite its dark appearance and apparent opacity to light. However, specific preparations or conditioning procedures are required to unveil these characteristics. Henceforth, coal has the potential to be scrutinized under visible light through the utilization of either transmittance or reflectance techniques. Transmittance is a metric that quantifies the degree of light absorption across different wavelengths. This parameter can be ascertained for thin coal sections, compressed coal particles in a potassium bromide disk, coal extract solutions in solvents such as pyridine, or coal films generated via evaporation of a dispersing liquid.

The examination of reflectance, particularly in coal, holds considerable importance due to its ability to offer valuable insights into several crucial aspects of coal, such as the maceral structure. This, in turn, facilitates the prediction of processing behavior (Davis, 1978; Davis et al., 1991). The term "coal reflectance" pertains to the degree of reflection of a polarized light beam that is directed towards a polished surface of coal, which has been prepared using a specific methodology. The aforementioned procedure produces specimens that are employed for microscopic assessment of the reflectivity of organic constituents present in a processed coal sample, in addition to ascertaining the proportion of physical components of coal in terms of volume.

The coal undergoes a process of comminution to attain a particle size that enables it to traverse a screen of mesh number 20 while generating negligible quantities of fine particles. Subsequently, the resultant fragments are consolidated into compact briquettes, which are bound by a cohesive agent. The briquette is subjected to a polishing process wherein one of its surfaces is refined using increasingly fine abrasives until a surface that is devoid of scratches and char is attained. The utilization of a metallurgical microscope, commonly referred to as an opaque-ore microscope, is used for the purpose of evaluating the reflectance of coal when subjected to vertical illumination through the use of polarized light.

The typical testing method for determining vitrinite reflectance entails the microscopic analysis of polished vitrinite, and other macerals present in coals across the spectrum of lignite to anthracite. This analysis aims to establish the average maximum reflectance and average random reflectance of the macerals, both measured in oil. The identical testing methodology can be utilized to quantify the reflectivity of additional macerals. The current methodology entails the utilization of a microscopic system to visually identify the amount of light reflected from the greasy surface of a coal specimen, which is illuminated in immersion oil. The present methodology facilitates the computation of the reflectivity of vitrinite. Upon completion of the calibration process for photometric equipment, the reflected light is evaluated in comparison to established reflectance standards, and the results are recorded as a percentage of reflectance.

The experimental technique used to determine the mean maximum reflectance of the vitrinite component in coal is frequently utilized as a rank indicator for coal characterization as a feedstock for various processes such as carbonization, gasification, liquefaction, and combustion, regardless of its petrographic composition. In order to utilize mean maximum reflectance as a ranking parameter, it is imperative to furnish the identified categories of vitrinite.

Upon the application of a commercially supplied immersion oil onto the surface of the sample, a series of measurements are conducted to ascertain the highest level of reflectance exhibited by the relevant coal constituent, such as vitrinite. These values are subsequently compared to measurements obtained from high-index glass standards, which encompass reflectance values ranging from 0.302% to 1.815%.

Therefore, despite the fact that coal usually seems to be a dark mass, it can exhibit a wide range of hues when sliced thinly or polished. For instance, fusinite and micrinite appear white when illuminated by incident light, whereas exinite displays yellow color. Contrarily, exinite displays an orange color when illuminated by transmitted light. It should come as no surprise that these color distinctions are utilized in the process of distinguishing the various types of macerals. Additionally, the reflectance of coal changes with the amount of carbon it contains (as shown in Figure 5.6), and the reflectance data acquired using air are significantly greater than those collected with oil as the medium.

Coal's volume percentage of its physical components is used to help characterize it for usage in combustion, gasification, and carbonization processes. In order to measure the volume percent of the components in an appropriate pulverized coal sample, microscopic analysis is conducted based on their reflectance, optical characteristics, and morphology. By observing an adequate number of points and identifying the characteristic indicators of each component, the percentage of each component can be calculated. The surface analysis of a sample involves the determination of minute component areas exclusively. Nonetheless, in cases where the sections are distributed haphazardly within the specimen, the ratios of both area and volume exhibit constancy.

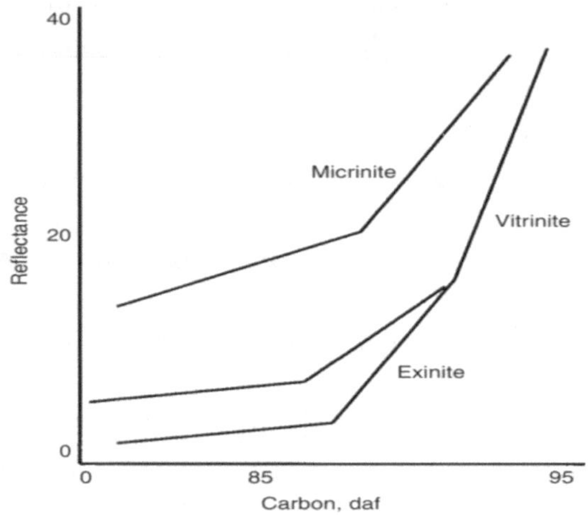

Figure 5.6. *The reflectance of a material exhibits variation in response to changes in its carbon content.*

Source Dorman, Creative Commons License.

5.4. REFRACTIVE INDEX

The determination of the refractive index of coal can be accomplished through a comparative analysis of its reflectance in air and cedar oil. As a component of a routine testing protocol, the lustrous vitrinite surfaces and other macerals present in coal specimens spanning the lignite to anthracite rank spectrum are assessed for their average maximum reflectance and oil-based random reflectance. The reflectance of additional macerals may be assessed utilizing an identical testing methodology. The refractive index values of vitrinite, which is a coal group, typically fall within the range of 1.68 to 2.02.

5.5. CONDUCTIVITY

The resistance of a block of coal with a cross area of 1 cm^2 and a length of 1 cm is used to describe the electrical conductivity of coal and is represented by the unit p (ohm-centimeters). Materials between these limitations are classed as semiconductors. Materials are categorized as insulators if their specific resistance is more than or equal to 1 × 1015 Ωcm and as conductors if it is less than or equal to 1 Ωcm. Temperature, pressure, and the coal's moisture content are only a few of the variables that affect electrical conductivity. When coal is heated to high temperatures, especially those exceeding 600°C, the structure of the coal starts to degrade, which greatly increases electrical conductivity. Electrical conductivity is significantly affected by moisture, leading to an increase in conductivity. The coal is typically kept in a dry, oxygen-free atmosphere to avoid any anomalies from the conductance caused by water, and to reduce potential issues, especially due to the existence of water. Initial measurements are typically taken at about 200°C and then kept at lower temperatures. Anthracite is a semiconductor with a specific resistance range of 1 to 104 cm, but bituminous coal has a range of 105 to 1,012 cm.

Subbituminous coal also exhibits semiconductor properties. Coals with carbon content ranging from 80% to 92% demonstrate the highest resistances, rendering them akin to virtual insulators. Hence, the coal's ability to conduct electricity, denoting the movement of electric charge, is commonly represented by its electrical resistivity. In order to ensure the attainment of values that are representative, the utilization of specific electrical resistance and specific conductance is implemented. The first term pertains to the opposition to the flow of electric current offered by a cube with unit dimensions quantified in terms of ohm-unit length.

On the other hand, the second term denotes the inverse of the first term. When analyzing resistivity or conductivity, it is recommended to employ suitable instrumentation, such as a fluxmeter, an ammeter-voltmeter system, a Wheatstone bridge, and electrometers for direct-current insulation resistance determination. These instruments are recommended for conducting transient resistivity measurements. This particular testing procedure is suitable for direct-current operations that aim to assess various properties of electrical insulating materials, including dc insulation resistance, volume resistance, volume resistivity, surface resistance, surface resistivity, as well as conductance and conductivities that correspond to these properties. Nonetheless, this approach proves to be inadequate in evaluating the electrical resistivity or conductivity of materials with moderate conductivity, thereby necessitating the adoption of an alternative technique.

This method is used to determine the electrical resistance or conductance of materials that fall under the category of moderately conductive substances, which are neither excellent electrical insulators nor proficient conductors. This approach is applicable to any materials with a volume resistivity ranging from 1 to 10 ▲ cm or a surface resistivity ranging from 10 to 10 ▲ (per square). The intended temperature and relative humidity for measurements are 23°C and 50%, respectively. However, the underlying principles of this method can be applied to specimens examined at lower or higher temperatures and relative humidity levels.

All electrical valuations rely heavily on water content; the dielectric constant for water is roughly 81, and the resistivity is approximately 106 ▲ cm. The dielectric constant has been employed to determine the moisture content of coal, as per Speight (1994). However, due to the distinct electrical characteristics of chemically and physically bound water, the effect does not qualify as an additive. Electrical conductivity and dielectric constant increase as moisture content increases, whereas resistance and dielectric strength decrease. As a result, save for special applications (such as measuring the dielectric strength of subterranean coal blocks); electrical tests necessitate precise drying of coal before conducting experiments.

When considering the acceptability of coal for specific applications, the features indicated in this chapter must always be kept in mind. It should also be remembered that coal that appears unfit may become excellently suited after a practical pre-treatment process, according to Speight (1999).

The electrical conductivity of coal can be attributed in part to the partial mobility of electrons within the structural lattice of coal, which is influenced by unpaired electrons or "free radicals." Additionally, the presence of mineral matter in coal can impact its electrical conductivity. Notably, the conductivities of different coal macerals exhibit significant variations, with fusains demonstrating superior electrical conductivity compared to durain, vitrain, and clarain.

5.6. DIELECTRIC CONSTANT

The dielectric constant gauges coal's potential to store electricity as well as its electrostatic polarizability. When describing coal, the dielectric constant, a measurement of the dielectric coal's electrostatic polarizability, is more helpful than electrical conductivity. The dielectric constant of coal is assumed to be connected to the polarizability of the electrons in the clusters of aromatic molecules in its chemical structure.

There is a large frequency range for which experimental approaches are applicable. Commercially available Q-meters, dielectric constant meters, and other instruments are used for high-frequency measurements;

for low frequencies, the impedance bridge approach is frequently used. For the purposes of comparing and interpreting results, the magnitudes of the frequencies that were experimentally applied are crucial.

The coal's dielectric constant is highly relied upon its rank (van Krevelen, 1961; Speight, 1994). The minimum value of the dielectric constant for dry coals is 3.5, and this value is reached at approximately 88% weight-to-weight carbon content in the bituminous coal range. Both anthracite and lignite have a dielectric constant that abruptly approaches 5.0 as they progress from lower to higher carbon content. In most cases, even for thoroughly dried-out coal, there is a notable difference from the Maxwell relation, which states that the dielectric constant equals the square of the refractive index for polar non-conductors.

Both conductivity and dielectric constant are quite sensitive to the amount of water present in the system. In point of fact, the dielectric constant may be utilized as a method for finding out the amount of moisture present in coal. Measurements of coal's dielectric constant that are meaningful require the coal to be dried to a constant dielectric constant first, and dielectric constant measurements are performed on numerous different types of coal. These can be blocks of coal that have been precisely formed, mulls of coal that have been suspended in solvents with a low dielectric constant, or chunks of powdered coal that have been embedded in a paraffin matrix.

The dielectric constant shifts depending on the rank of the coal (Chatterjee & Misra, 1989). The theory stating that the dielectric constant is equal to the square of the refractive index holds true only for coal, which has the lowest possible dielectric constant. This rule, which applies to all non-conducting and non-polar substances, is true. It is possible that the loss of polar functional groups is the cause of the lowering value of the dielectric constant with increasing rank; however, the significance of the existence of polarizable electrons is not completely understood. It also seems that the presence of naturally occurring water in coal has a significant impact on the dielectric characteristics (Chatterjee & Misra, 1989).

5.7. DIELECTRIC STRENGTH

The measure of dielectric strength is expressed in volts per unit thickness of the sample and denotes the voltage level at which dielectric breakdown occurs. It is frequently assessed at frequencies used in commercial power systems. The present methodology outlines procedures for assessing the dielectric potency of solid insulating substances at frequencies relevant to electricity generation. However, all tests are conducted at a frequency of 60 hertz. The testing methodology under consideration is amenable to implementation across a range of frequencies spanning from 25 Hz to 800 Hz. However, it is important to note that frequencies exceeding 800 Hz may produce dielectric heating. The principal objective of this technique is to identify the voltage at which dielectric breakdown occurs in a specimen throughout its thickness.

Empirical data robustly show that bituminous and anthracite coal exhibit electrical anisotropy. Specimens arranged perpendicular to the bedding plane have higher resistivity/lower conductivity than those oriented parallel to the plane.

5.8. SPECIFIC RESISTANCE

The electrical resistance of a unit cross-section area and length of an object is called its Specific resistance or resistivity. Its unit is ohm-centimeters and is expressed as follows:

$$p = \frac{RA}{L}$$

where; *p* denotes specific resistance; *R* denotes substance resistance; *A* denotes a cross-sectional area; and *L* denotes length. Following the direction of measurement, the coal's specific resistance can range from thousands to millions of ohm centimeters.

When wet coal is subjected to elevated temperatures ranging from 0 to 200°C, its electrical resistivity increases (with a contemporary reduction in dielectric constant). This is because of moisture loss. A temperature increases after moisture removal results in reduced resistance (and a greater dielectric constant). As with any semiconductor, the dependence of conductive characteristics on temperature is mostly exponential. The influence of temperature on electrical characteristics is transformable at lower temperatures. The beginning of transformable effects depends on rank and begins at 200 to 400°C for bituminous coal and 500 to 700°C for anthracite.

It is not completely known how the presence of volatile materials can affect electrical characteristics. While some data suggest that there is no special association, other data suggest that there is a minor decrease in resistivity associated with a reduction in the amount of volatile materials. Both the kind of volatile stuff and the amount of it can have an effect on the electrical characteristics. It is important to note, however, that even samples with the same amount of volatile matter can display distinct features due to the diverse sources of the samples. Electrical characteristics can be affected by a material's ash concentration and mineral matter, with the former having a more noticeable effect).

5.9. SUMMARY

Physical and electrical properties of coal can vary depending on factors such as coal rank, composition, and impurities. Coal has a relatively high density and can have varying degrees of hardness, ranging from relatively soft to extremely hard. Coal can contain a significant amount of moisture, which affects its handling and transportation. Coal is generally a poor conductor of electricity, and the resistivity of coal refers to its opposition to the flow of electric current and exhibits high resistivity. Coal has a relatively low dielectric constant. It is important to note that the physical and electrical properties of coal can vary significantly depending on the specific type and quality of coal.

REFERENCES

1. Agrawal, P. L., (1959). *Proceedings of the Symposium on the Nature of Coal* (p. 121). Central Fuel Research Institute, Jealgora, India.
2. Baughman, G. L., (1978). *Synthetic Fuels Data Handbook*. CameronEngineers, Denver, CO.
3. Berkowitz, N., (1979). *An Introduction to Coal Technology*. Academic Press, San Diego, CA.
4. Braunstein, H. M., Copenhauer, E. D., & Pfuderer, H. A., (1977). *Environmental Health and Control Aspects of Coal Conversion: An Information Review*. Report ORNL-EIS-94. Oak Ridge National Laboratory, Oak Ridge, TN.
5. Chatterjee, I., & Misra, M., (1989). *Proc. Mater. Res. Symp. Mater. Res. Soc.*, 189, 195.
6. Davis, A., (1978). In: Karr, Jr. C., (ed.), *Analytical Methods for Coal and Coal Products* (Vol. 1). Academic Press, SanDiego, CA, Chapter 2.
7. Davis, A., Mitchell, G. D., Derbyshire, F. J., Rathbone, R. F., & Lin, R., (1991). *Fuel*, 70, 352.
8. Dorman, H. N. M., Huntjens, F. J., and van Krevelen, D. W., (1957). *Fuel*, *36*, 321.
9. Ergun, S., (1951). *Anal. Chem.*, *23*, 151.
10. Ettinger, I. L., & Zhupakhina, E. S., (1960). *Fuel*, *39*, 387.
11. Evans, D. G., & Allardice, D. J., (1978). In: Karr, Jr. C., (ed.), *Analytical Methods for Coal and Coal Products* (Vol. 1). Academic Press, SanDiego, CA, Chapter 3.
12. Gan, H., Nandi, S. P., & Walker, P. L. Jr., (1972). *Fuel*, *51*, 272.
13. Kalliat, M., Kwak, C. Y., & Schmidt, P. W., (1981). In: Blaustein, B. D., Bockrath, B. C., & Friedman, S., (eds.), *New Approaches in Coal Chemistry*. American Chemical Society, Washington, DC, Chapter 1.
14. King, J. G., & Wilkins, E. T., (1944). *Proceedings of the Conference on the Ultrafine Structure of Coal and Cokes* (p. 45). British Coal Utilization Research Association, London.
15. Kroger, C., & Badenecker, J., (1957). *Brennst. Chem.*, *38*, 82.
16. Labuschagne, B. C. J., (1987). *Proceedings of the 4th Annual Pittsburgh Coal Conference* (p. 417). University of Pittsburgh, Pittsburgh, PA.
17. Labuschagne, B. C. J., Wheelock, T. D., Guo, R. K., David, H. T., & Markuszewski, R., (1988). *Proceedings of the 5th Annual International*

Pittsburgh Coal Conference. University of Pittsburgh, Pittsburgh, PA, p. 417.

18. Lyman, W. J., Reehl, W. F., & Rosenblatt, D. H., (1990). *Handbook of Chemical Property Estimation Methods: Environmental Behavior of Organic Compounds*. McGraw-Hill, New York.
19. Mahajan, O. P., & Walker, P. L. Jr., (1978). In: Karr, Jr. C., (ed.), *Analytical Methods for Coal and Coal Products* (Vol. 1). Academic Press, San Diego, CA, Chapter 4.
20. Milewska-Duda, J., (1991). *Arch. Mineral. Sci.*, *36*, 369.
21. Ramesh, R., Francois, M., Somasundaran, P., & Cases, J. M., (1992). *Energy Fuels*, *6*, 239.
22. Speight, J. G., (1978). In: Karr, Jr. C., (ed.)., *Analytical Methods for Coal and Coal Products* (Vol. 2). Academic Press, San Diego, CA, Chapter 22.
23. Speight, J. G., (1994). *The Chemistry and Technology of Coal* (2nd edn.). Marcel Dekker, New York.
24. Speight, J. G., (1999). *The Chemistry and Technology of Petroleum* (3rd edn.). Marcel Dekker, New York.
25. Van Krevelen, D. W., (1957). *Coal: Aspects of Coal Constitution*. Elsevier, Amsterdam.
26. Walker, P. L., Jr., (1981). *Philos. Trans. R. Soc. (London)*, *300A*, 65.

CHAPTER

6

Mechanical and Thermal Properties of Coal

Contents

INTRODUCTION

Compared to proximate evaluation, final (elemental) analysis, and physical characteristics, the mechanical characteristics of coal received limited utilization and infrequent mention in scholarly publications. The exclusion of these properties is a significant oversight, as they hold significance and warrant consideration in forecasting the behavior of coal throughout mining, dealing with, and preparation processes (Ai, Xue, Wei, Dou, & Miao, 2018).

The mechanical characteristics of coal hold significant importance when forecasting its strength and conduct during mining operations, where the sturdiness of coal faces along with the strength of fossil fuels pillars, are critical variables. The mechanical characteristics of coal hold significant importance in various domains of coal technology, including coal extraction (for the planning and execution of cutting equipment), comminution (for choosing and construction of mills), storage (for the characteristics of flow or a shear failing), handling (for shatter while abrasion during transportation), and several other aspects (Honda, Sanada, & Furuta, 1966).

The mechanical characteristics of coal hold significant importance when forecasting its durability or conduct during mining operations, where the sturdiness of coal faces along with the strength of fossil fuels pillars, are critical variables. The mechanical characteristics of coal hold significant importance in various domains of coal technological advances, including coal extraction (for the planning and execution of cutting equipment), comminution (for choosing and construction of mills), storing (for flow characteristics and the shear failing), handling (for shatter while abrasion during transportation), and several other aspects (Kalaw et al., 2015).

6.1. MECHANICAL PROPERTIES

6.1.1. Strength

Various techniques for determining the comparative potency, fragility, or comminution capacity of coal involve the utilization of a ceramic jar mill, wherein each coal specimen is subjected to a specific number of revolutions. Screen examinations of the input and output materials approximate the extent of the fresh surface area generated. The grindability of coals is evaluated by measuring the quantity of fresh surface area generated during testing and contrasting it to that of reference coal (Li et al., 2016). The singular test technique that is currently accessible pertains to the evaluation of coke responsiveness and coke power after the reaction. The present study outlines the methodology employed to ascertain the reactivity of lump coke within the presence of carbon dioxide (CO2) at high temperatures, as well as its strength post-reaction in carbon dioxide gas, through specialized instruments and methods involving tumbling within a cylindrical chamber. The test procedure involves subjecting a predetermined quantity of dehydrated coke, with specific origin and dimensions, to a reaction with a certain amount of carbon dioxide gases within a retort, at a precisely controlled elevated temperature of 1,100 ± 5°C (2,012 ± 9°F) for a designated duration (Xie, Zhao, Liu, Zhang, & Xue, 2012). The reaction index (CRI) and coke energy after reaction (CSR) are ascertained using the reacting coke residue. The Coke responsiveness index is determined by the weight loss after a reaction. The coke power following reaction is determined by the weight following the screening of the dropped reacted coke, which is subjected to a specified number of changes at a chosen turning pace. The significance of the test method lies in its application to coke, as the descending coke clumps in the combustion chamber respond with opposition carbon dioxide and experience abrasion due to inter-particle friction and contact with the combustion walls. The simultaneous operations experience physical deterioration and chemical reactions with the coke aggregates, generating a surplus of small particles that may reduce the permeation of the weight and lead to elevated coke consumption and diminished output of the hot steel. The present test methodology has been devised to indirectly assess coke's performance within the blast furnace. The testing technique can be utilized to ascertain the potency of coal, subject to certain modifications that may be necessary based on the type of coal and the specific data sought (Czapliński & Hołda, 1982).

The compressive force of coal is a frequently studied aspect of its strength properties, as it is closely linked to the stability of coal pillars utilized for roof reinforcement in mining activities. Consequently, it is unsurprising that numerous tests have been developed to gauge the compressive property of coal. The approach employed for testing coal must be precise and rational, as the characteristics of the research inquiries and the specific coal being tested. Failure to do so may raise questions about the validity of the testing process (Lu, Zhang, Sa, & Si, 2019). Anthracite was subjected to compression tests in lenses with dimensions of 2 × 2 inches and 5 × 5 centimeters for the lowest point, or heights of one, two, and four inches. These prisms were carved parallel to the bed planes within the coal, and compression was applied perpendicular to these planes. The compressive strengths for prisms measuring 1 inch in height vary between 3,200 and 10,900 psi, with an average value of approximately 6,000 psi. Additionally, research has demonstrated that the breaking strength of coal exhibits an inverse relationship with the square root for the height during the specimen being tested (Jones, Bell, & Schraufnagel, 1988). The strength of bitumen coal specimens is influenced by their lateral dimensions, with smaller specimens exhibiting greater strength than larger ones (Table 6.1). This disparity may be attributed to fracture airplanes or cleats within larger specimens.

Table 6.1. Variation of Compressive Strength with Size

Size	Average Maximum Load
2.5–4-in. cubes	2,486
7–8-in. cubes	2,170
10–12-in. cubes	2,008
12 × 12 × 18 in. high	1,152
Approx. 30-in. cube	817
Approx. 54-in. cube	306

Source: Baughman, Creative Commons License.

The relationship between the compressive strength and coal position has been observed, and a graph depicting the strength as a function of volatile matter indicates that the typical minimum value is 20–25% dry, ash-free volatile matter (Song & Zhang, 2021). This trend is evident for both parallel and perpendicular enlargement concerning the bedding plane, as shown in Table 6.2 and Figure 6.1.

Table 6.2. Variation of Compressive Strength with Rank

Coal Rank	Compressive Strength (psi)
Anthracite	2,370
	3,390
	2,000
	1,740
Bituminous	310–2,490
	500
Lignite	2,190–6,560

Source: Hiorns, Creative Commons License.

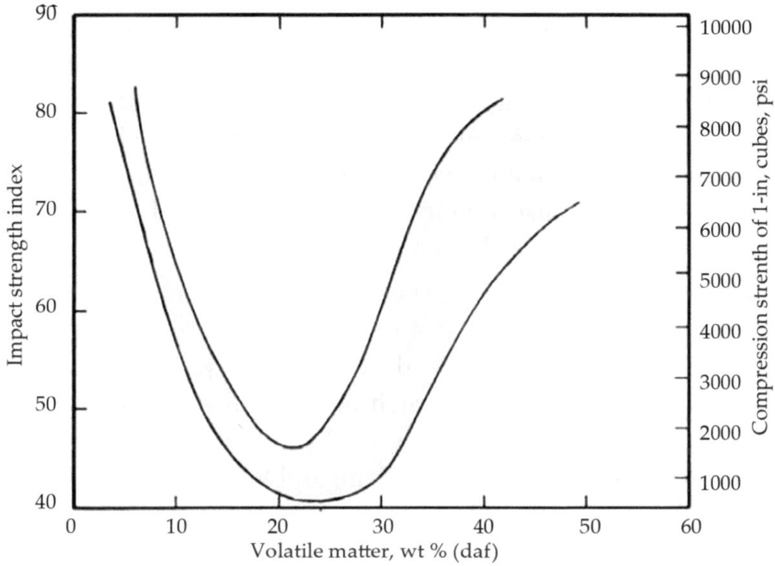

Figure 6.1. *Variation of coal strength with volatile matter production.*

Source: Brown, Creative Commons License.

6.1.2. Hardness

A significant amount of data is accessible regarding the characteristics of coal that pertain to its firmness, such as its susceptibility to crumbling and its ability to be ground. However, there needs to be more knowledge regarding the innate toughness of coal. Nonetheless, it is noteworthy that toughness can be deemed analogous to friability. In this regard, a technique has been established to assess the fragility of coal, which closely resembles the technique employed to gauge the comparative degree of Coke's durability against degradation. The present study outlines a methodology for acquiring a comparative assessment of the durability of coke against impact and abrasion-induced deterioration. The prescribed procedure entails subjecting a designated-sized, desiccated Coke sample to a rotational motion within a drum for a predetermined number of changes during an established pace. The stability and hardness factors associated with coke are established through sieve testing after its therapy (Tiryaki, 2005).

The present testing methodology may be implemented for coal samples while bearing in mind that the original purpose that drove the method was developed, and the results obtained may be relevant to the determination of coal hardness. The utilization of testing on coal, which is typically less dense than coke, may necessitate modifications to the experimental parameters. Nevertheless, the information obtained can be valuable in assessing the coal's hardness and potential deterioration during transportation and manipulation. The procedures above entail exposing the coke to stress and scratching, and the assessment technique utilized is a comparative gauge of the coal's durability against fragmentation when exposed to these forms of deterioration (Klawitter, Esterle, & Collins, 2015).

Determining coal's scratch resistance entails measuring the load exerted on a pyramid-shaped steel point that is necessary to produce a scratch with a width of 100 μm upon the polished exterior of a specimen. According to Table 6.3, anthracite exhibits a scratch hardness that is roughly six times greater than soft fossil fuels, while pyrite displays a hardness that is nearly 20 times greater than that of soft coal. Comparable findings were observed in the case of anthracite along with cannel coal. However, commonly believed to be the most rigid constituent of coal, it exhibited only a marginally higher level of hardness than train-along with cannel coal (Szwilski, 1985).

While the ability to resist coals towards abrasion may not seem commercially significant, the coal's roughness is crucial. The abrasion of grinding components caused by coal leads to expenses related to maintenance, which represents a significant portion of the overall cost of pulverizing coal for use as fuel. In addition, it is imperative to consider the varying degrees of abrasiveness exhibited by different types of coal during the selection process for pulverized fuel plants (Yaralı, 2017).

Table 6.3. Representation of Scratch Hardness of Coal

Material	**Scratch Hardness Relative to Barnsley Soft Coal**
Anthracite, Great Mountain	1.70
Anthracite, Red Vein	1.75
Welsh steam	0.29
Barnsley hards	0.85
Barnsley softs	1.00
Illinois coal	1.10
Cannel	0.92
Carbonaceous shale	0.69
Shale	0.32

Pyrite	5.71
Calcite	1.92

Source: Berenbaum, Creative Commons License.

6.1.3. Friability

The capacity of coal to endure size reduction during handling is considered a significant indicator of its strength. The friction test is a commonly used method for assessing coal power; various factors, including toughness, flexibility, fracture features, and strength, influence its propensity towards breakage upon handling (known as friability) (Yancey & Zane, 1933).

The concept of friability is significant because coals with higher friability tend to produce lesser quantities of coarse sizes, which may or may not be preferable depending on their intended application. Additionally, friable fuel may exhibit a greater surface area. The surface above facilitates a quicker oxidation process, thereby creating a more conducive environment for unplanned ignition, degradation in coking characteristics of coking coal, and other concomitant alterations that occur during oxidative stress. The economic factors associated with the flexibility of coal have served as a driving force for the advancement of laboratory-based friability testing (Falcon & Falcon, 1987).

The tumbling test is a method employed to assess the friability of coal, specifically about a given size of calibrated coal. This method quantifies the propensity of coal to fracture into smaller fragments under repetitive handling conditions, whether at the mining site, during distribution, or through the end user. The testing mentioned methodology is deemed advantageous in assessing the comparability of coal samples about their friability, as opposed to determining principles within a restricted range to underscore their dissimilarity (Hertzog & Cudworth, 1938). This particular testing approach can also potentially indicate the comparative degree to which coals of different sizes may experience size deterioration in specific mechanical feeding apparatuses. The present test methodology can be employed to distinguish between various levels and categories of coal, thereby rendering it a valuable tool for coal categorization purposes (Hauserman & Tye, 1984).

The employed technique utilizes a cylindrical mill made of porcelain jars equipped with all three lifters to help in the tumbling of the coal. A predetermined quantity (typically 1,000 g) of coal that has been sized is subjected to tumbling within a mill for the designated duration and at a specific rate of revolutions each minute. Subsequently, the coal is extracted and subjected to screening procedures, whereby the degree of friability is quantified as the percentage decrease in the mean particle dimensions throughout the experimentation. If the mean particle dimensions of the crashed coal were 75% of the initial sample, the degree of friability would amount to 25%. Similar to various other assessments, the test parameters could be modified to align with the objective of the inquiry. However, it is imperative to document the exact parameters alongside the data (Greenhalgh & Emerson, 1986).

Efforts have been undertaken to establish a correlation between coal friability and rank, as presented in Table 6.4. It was observed that lignite with high moisture content exhibited the lowest degree of friability, while the friability for coal raised with its rank, reaching a peak for low-volatile bitumen coal. The degree of flexibility anthracites is similar to the characteristics of sub-bituminous coal. Compared to bituminous coal, anthracite, and coal exhibit greater strength and are considerably more resilient to breakage. This contrasts with certain semi-bituminous coals that display a high degree of friability (Shkoller, Kazimirov, Temlyantsev, & Basegskiy, 2015).

Table 6.4. Variation of Friability with Rank

Coal Rank	**Number of Tests**	**Friability (%)**
Anthracite	36	33
Bituminous (low-volatile)	27	70
Bituminous	87	43
Subbituminous A	40	30
Subbituminous B	29	20
Lignite	16	12

Source: Yancy, Creative Commons License.

The correlation between coal friability and rank is relevant to its proclivity for spontaneous combustion. Coals with high rank, which are friable and low in volatility, exhibit low susceptibility to oxidation despite excessive particles and bigger surfaces resulting from handling. Coal using lower ranks tends to oxidize more easily. It is typically less friable, meaning it is more resistant to destruction in size and the subsequent increase in surface area that results from oxidation. The foremost contributor to instability in coal stockpiles is undoubtedly oxidation caused by oxygen from the atmosphere.

Any additional factors, including friability, intensify the main oxidation impact (Aida & Okamoto, 1960).

6.1.4. Grindability

The concept of grindability pertains to the comparative facility with which a given coal sample can be crushed about selected typical coals. The Hardgrove equipment has widely adopted the Hardgrove procedure as the established technique for assessing coal grindability. An alternative test procedure exists for evaluating the Hardgrove grindability ranking of oil coke, which may be deemed appropriate for exceptionally hard coal in cases where the conventional test method fails to yield satisfactory levels of replication and repeatability. It may be necessary to make appropriate alterations to the testing methodology for petroleum coke (Sengupta, 2002).

The technique of the Hardgrove grindability examination pertains to assessing the comparative ease of crushing and the grindability of coal about standard coal. The process involves subjecting a prepped and appropriately sized sample to a specific amount of crushing energy using a miniature pulverizer. The resulting size distribution of the crushed product is then determined through sieving. The size distribution that ensues is utilized to generate an indicator that pertains to the difficulty associated with the grinding process (Deniz & Umucu, 2013).

The physical properties of high-volatile bitumen, coal, and lignite may be altered due to the release of natural and seam humidity during handling and getting ready. The modification frequently results in a modification of the grindability properties observed during laboratory testing, potentially leading to varying indices based on factors such as drying conditions and moisture content. This variability can introduce discrepancies affecting fossil fuel data's reliability and reproducibility (Tiryaki, 2005).

The grindability assessment procedure involves employing a ball-and-ring mill configuration, whereby a sample of coal with minimal variation in size is subjected to grinding for a predetermined number of revolutions, typically 60 (Miroshnichenko, Desna, Koval, & Fatenko, 2019). Subsequently, the resulting ground goods are subjected to sieving, and the grindability indices are determined by referencing a calibration chart (Figure 6.2) that correlates the amount of undersized material produced with the index value).

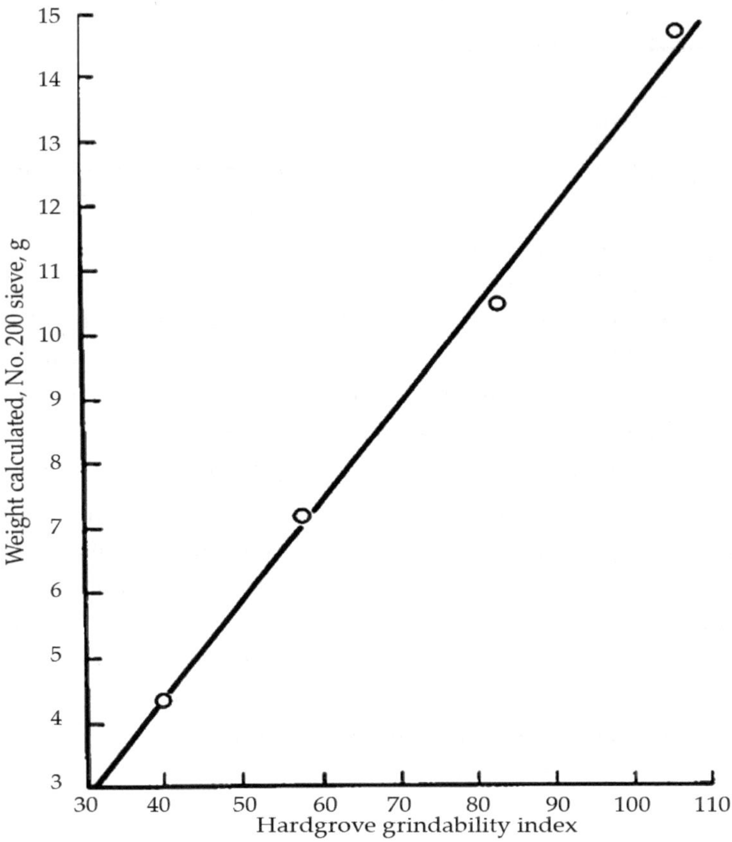

Figure 6.2. *Representation of calibration chart for Hardgrove grindability index.*

Source: Henke, Creative Commons License.

Figure 6.3 illustrates a correlation between coal ranking and grindability, indicating that coals with lower volatility and medium rank are generally simpler to grind. However, it should be noted that the relationship between rank and grindability is not precise enough to estimate grindability based solely on fossil fuels analysis. Low-volatile bituminous coal typically demonstrates a grindability index that exceeds 100, while high-volatile fossil fuels show grindability indices ranging from roughly 55 to 35 to 40. Nonetheless, a connection exists between flexibility and grindability, given that coal with a soft and readily breakable structure typically demonstrates a comparatively elevated grindability index (Gil, García, Pevida, & Rubiera, 2015).

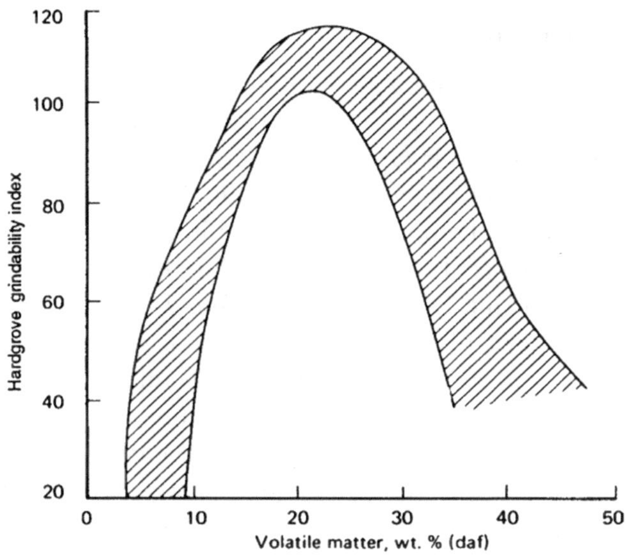

Figure 6.3. *Variation of Hardgrove grindability index with volatile matter.*

Source: Berkowitz, Creative Commons License.

6.1.5. Dustiness of Index

Mitigating dirt in coal-getting-ready plants and underground mines is a crucial safety consideration, so ongoing efforts are to enhance dust removal methodologies. Efforts have been made to enhance the certainty of coal behavior under specific circumstances, particularly regarding the generation of dust, as advancements in methods have been made (Mikula & Parsons, 1991).

The dustiness index is a metric utilized to ascertain the comparative measures that indicate the quantity of particulate matter generated while handling coal in a standardized fashion. A sample of coal weighing 50 pounds is positioned onto a slide plate within a metal cabinet of predetermined dimensions (Hamelmann & Schmidt, 2003). Upon removal of the plate, the specimen descends into a compartment; after that, two slides are introduced into the receptacle after 5 seconds. The slides are designed to accumulate particulate matter suspended within the air for either 2 minutes (in the case of coarse dust) or 10 minutes (in the case of fine dust). The dustiness indices are quantified as a multiple of 40 times the weight of dust particles that have settled within 2 minutes or after an additional eight-minute interval (Anderson, Scales, & Ralston, 1986).

6.1.6. Cleat Structure

Coal deposits frequently exhibit a system for joint planes whereby the joint planes, also known as cleats, are typically oriented parallel with the bed planes. Vertical orientation is typically preferred for cleat joints. The facial cleat is the primary jointing system, while the butt settled a cross-system for jointing. Moreover, the presence of shoes in coal significantly impacts the characteristics of a coal reservoir. According to research, drilling holes through coal parallel to the face cleat has been found to produce methane gases from the deposit at a rate of 2.5, with 10 times higher than holes dug parallel with the butt cleat (Laubach, Marrett, Olson, & Scott, 1998). The dimension of run-of-mine (ROM) fossil fuels may be determined by the settled system that causes a fracture or the regularity of cleats. Typically, a set of cleats will exhibit a perpendicular orientation of approximately 90° (Mostaghimi et al., 2017). The cutting components' configuration significantly impacts the coal excavation machinery's performance, as depicted in Figure 6.4).

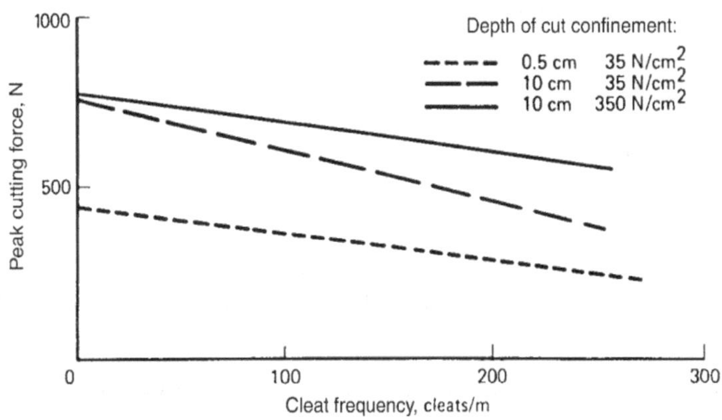

Figure 6.4. *Effect of cleats on the cutting force.*

Source: Baughman, Creative Commons License.

6.1.7. Deformation and Flow Under Stress

Scholars have researched the theological properties of coal, specifically how it deforms and flows, to apply this knowledge to the characterization of coal in mining contexts. Additionally, the elasticity of coal, which refers to its ability to recover its original shape after undergoing deformation according to strain, has been the subject of investigation. Obtaining significant measures of coal flexibility can be a challenging task (Zhang, Liu, & Elsworth, 2008).

The elimination process, typically accomplished through a wheel of grinding containing embedded grit within a rubber matrix, generates heat that can induce plastic distortion within the coal surface, ultimately impacting its plastic properties. The utilization of water for cooling the grinding connect and transporting particles has the potential to be assimilated by the coal, thereby influencing its elastic properties. Moreover, the presence of discontinuities within the coal structure results in a significant degree of variation between samples (Zhu, Liu, Sheng, & Elsworth, 2007).

6.2. THERMAL PROPERTIES

6.2.1. Calorific Value

The calorie content refers to the amount of heat generated through the combustion of a specific amount of fossil fuels in a bomb calorimeter while being exposed to oxygen and a predetermined set of circumstances. The determination of the calorie content of coal is carried out through the utilization of a bomb calorimeter, which works by employing either a stationary (isothermal) or an adiabatic technique. The net calorific appreciation is of interest; an adjustment is applied. The unit of measurement utilized is energy per gram, and it is feasible to convert this unit to alternative units, such as 1.0 kilocalories per kilogram is equivalent to 1.8 British thermal units per pound or 4.187 kilojoules per kilogram (Ren et al., 2020).

The two commonly used measures of calorific value are the gross calorific value (GCV), higher heating value (HHV), calorific value (NCV), or the net lower calorific value (LHV). The distinction between the total calorie content and the actual calorific admire is attributed to the heat latent of water condensing into a vapor that arises from combustion. The total calorie content assumes that all vapor generated throughout the combustion process condenses completely. The net calorie content presupposes the exclusion of complete condensation of water from the combustion products. To standardize the impact of various factors, it is recommended to evaluate the calorific nutritional value of coal using the net calorific worth basis. The variability of coal's calorific value is contingent upon its ash and moisture content and specific type. Conversely, fuel oils exhibit more consistency in their calorific value (Tang, Wang, Lu, & Yang, 2015).

Thermal coal's calorie content, or heat of combustion, is a significant characteristic. The gross calorie value (CV) is a useful metric for determining the overall calorific amount of coal or coke, which can be applied in payment calculations. The utilization of this method is applicable in the computation of the calorific value of the sulfur content of coal, which is significant in verifying compliance with the regulatory standards for industrial fuels. The efficacy of processing processes may be assessed using the gross calorie content (GCV). The classification of coal can be determined by obtaining the gross energy content (Deng, Li, Xiao, & Shu, 2017).

The beneficial thermal value (UHV) is a measure of the energy density of coal. It can be calculated for noncaking coal using a formula considering the ash and moisture levels.

UHV (kcal/kg) = 8,900 – 138 × [ash content (wt.%) + Moisture content (wt.%)]

The calorie content of coal is not included when performing the closest or final analysis. It is considered a physical property of coal and is typically discussed in sections about its physical characteristics. Within the current context, the significance of the calorie content as a method for assessing coal necessitates its inclusion within this specific section and within the section that outlines the overall thermal characteristics of coal. Three general types of calorimeters can be distinguished to determine the calorie content of coal: The three types of calorimeters commonly utilized are: (i) those that employ solid oxidizing substances regardless of an open or a shut structure; (ii) those that utilize gaseous oxygen at roughly atmospheric pressure within a system that is open; and (iii) those that employ gaseous oxygen at high pressure within a system that is closed. All three categories of calorimeters are frequently utilized; however, solely those categorized under the third category can be deemed dependable for a moderately high level of precision (Van Krevelen, 1993).

When coal is combusted on an exposed grate under typical atmospheric pressures, negligible amounts of nitrogen dioxide are generated through the reaction of oxygen from the atmosphere with the nitrogen, the fuel, or the air. However, when subjected to the elevated temperature and pressure circumstances throughout the bomb, a greater quantity of nitrogen oxides (NOx) is produced, ultimately forming a solution that includes nitric acid through dissolution in water. The exothermic processes involving the combustion of nitrogen and subsequent dissolution of the resulting products in water necessitate the consideration of the heat released during the creation of nitric acid (Wen, Jian, & Wang, 2017).

Analogously, the combustion of sulfur or pyrites in the presence of air results in the oxidation of merely 5% of the sulfur with sulfur trioxide, while the remaining sulfur produces sulfur dioxide. However, in the bomb, nearly all of the sulfur undergoes combustion to form sulfur trioxide, which disintegrates in water to generate sulfuric acid. Consequently, it is imperative to rectify the amount of heat released during the combustion process for sulfur dioxide for trioxide, the dissolution of the trioxide within water resulting in sulfuric acid, along with the heat generated during the process of diluting of this acidic substance within the water (Bilgen & Kaygusuz, 2008).

Incorporating a correction factor to account for heat losses via radiation generated by the calorimeter is necessary when analyzing the observed temperature increase. A complex formula denotes rectifying this error; however, if the temperature data is systematically arranged on the laboratory sheet, the formula application becomes effortless and minimizes the probability of committing mathematical mistakes (Nguyen, Bui, & Bui, 2021).

It is recommended to calibrate the Beckmann thermometer used with the bomb calorimeter to account for its typical immersion depth during operation. To encompass the typical spectrum of laboratory temperatures, it is recommended that calibration be conducted for three distinct situations of the scale's zero point. Optimal values for these settings include 10°C, 15°C, and 20°C. Implementing a sequence of calibrations enables the automatic compensation of stem corrections that may arise and variations in the degree value on the thermometer's scale due to differing quantities of mercury in the bulb. This is in addition to any inherent differences within the dimension of a capillary carrier (Patel et al., 2007).

The full combustion of black may pose a challenge; however, introducing a precise quantity of liquid paraffin with an established calorific value can aid in the process. Raising the oxygen stress to 30 atm is deemed advantageous (Bilgen, Keleş, & Kaygusuz, 2012).

When determining the calorie content using a calorimeter for bombs, it is common practice to adjust the reported figure for the elevated temperature of sulfuric acid's creation and its dissolution by water. However, suppose the calorific value needs to be stated upon a dry, mineral-matter-free foundation. In that case, an additional adjustment must be made to the air-dried figure to account for the heat of the creation of ferric oxide resulting from the pyrites in the coal. Burning pyrite (FeS_2) generates heat at a rate of roughly 3,000 calories for each gram of pyrite sulfur. The appropriate correction factor depends on whether or not the calorie content is denoted as the heat produced per gram or pound (Majumder, Jain, Banerjee, & Barnwal, 2008). Table 6.5 reports different coals' energy, also known as heat content. However, it has garnered comparatively less focus than the calorie content. Variations in thermal evaluation have been reported as a method for determining the calorie content of coal.

Table 6.5. Variation of the Heat Content of Coal with Temperature

Temperature		**Heat Content**		
	As Tested		**Ash-Free Basis**	
°C	**°F**	**cal/g**	**Btu/lb**	**cal/g**
Lignite (Texas)				
32.7	90.9	11.8	21.2	13.5
69.3	156.7	20.2	36.4	22.5
95.3	203.5	25.4	45.7	27.7
34.4	273.9	39.2	70.6	42.5
Subbituminous B (Wyoming)				
42.3	108.1	14.1	25.4	14.5
65.0	149.0	19.4	34.9	19.8
89.7	193.5	26.4	47.5	26.9
112.6	234.7	34.0	61.2	34.6

Source: GregTomson, Creative Commons License.

6.2.1.1. Determination of Calorific Value

The conventional approach for ascertaining the total calorie content of coal involves the utilization of both an adiabatic calorie counter and a thermal explosives calorimeter. The techniques above involve the combustion of a sample of known weight within an oxygen bomb, which is subsequently immersed in a container filled with water and enclosed by a surrounding jacket (Muñoz-Guillena et al., 1992).

The adiabatic calorie counter is a thermometer designed to maintain zero heat head by adjusting the jacket temperature to match the calorimeter temperature. The methodology involves the combustion of a coal sample within the calorimeter while simultaneously adjusting the jacket temperatures to closely approximate the temperature of the water within the calorimeter. The calorific value is determined through the analysis of pre-and post-combustion observations. The isothermal technique involves igniting precisely measured coal samples within oxygen under regulated conditions to determine their calorie content. The temperature readings taken before, during, and after the combustion process are utilized to calculate the CV, given due consideration to the heat generated by other processes. The energy content of coal is typically denoted in British thermal units for each pound, kilocalories for each kilogram, or joules per kilogram. Notably, 1.8 Btu/lb equals 1.0 kcal/kg, equivalent to 4.187 kJ/kg (Lu et al., 2021).

According to ASTM D-3286, the isothermal calorie structure involves the correction of the temperature increase of the measuring instrument's water for any heat loss or gain that may occur due to the environment's jacket throughout the combustion of the specimen in question. The total heat generated during the combustion of the sample can be determined by multiplying the calorimeter's energy value with the fixed temperature increase in the two systems. The determination of the energy comparable, also known as the water comparable or the thermal capacity of a calorimeter, is achieved through the combustion of standardized specimens of benzoic acid that have been prepared (Patel et al., 2007).

After the detonation event, the bomb's constituents are transferred into a receptacle and subjected to titration utilizing a standardized sodium carbonate solution, thereby facilitating the quantification of the acid content (namely, nitric acid, HNO_3 along with sulfuric acid, H_2SO_4) generated during the combustion process. The large calorific value of solid fuel is determined by adjusting the entire heat generated in the calorimeter, which involves adjusting the number of acids, the quantity of fuse wire utilized during firing, and the level of sulfur in the specimen. This is achieved by multiplying the energy equivalent by the rectified temperature rise (Kök & Keskin, 2001).

An inherent limitation of this approach pertains to the accuracy of the temperature measurements. If a thermometer made from mercury in glass is employed, it must be set up precisely, and readings must be taken consistently. Numerous calorimeters are outfitted with digital thermometers that feature thermistor instruments and processors for regulating the ignition and documenting the temperatures at predetermined intervals. This mitigates the majority of the potential for human error in documenting alterations in temperature (Küçükbayrak et al., 1991). Initiating combustion in a coal sample within the oxygen bomb apparatus may present challenges.

Introducing oxygen too rapidly may cause the sample to be expelled from the crucible. Compacting the coal sample through a pellet could mitigate the risk of sample ejection. Coal possessing a significant amount of minerals, exhibits resistance to ignition. It may be advantageous to incorporate a predetermined quantity of ordinary aromatic acid into the sample and pelletize the mixture to address this issue (Altun, Hicyilmaz, & Kök, 2001).

Swiftly restoring the atmosphere of a bomb for atmospheric levels after its discharge may lead to the depletion of oxides of nitrogen and sulfur. It is necessary to adjust the total calorie content of the quantities of acid-forming metal oxides generated during the combustion process within the bomb. The loss incurred yields a significant calorific value. To prevent undesired consequences, it is imperative to restore the stress of the bomb gradually (Mason & Gandhi, 1983).

Regular inspection of the calorimeter devices is necessary to detect any alterations in the calorimeter's power equivalence, corrosion or impairment of the calorimeter container, harm to the breathable air bomb (even if it is minor), and any problems in the shakers, electrical wiring, or other components of the calorimeter. Any alterations or malfunctions have the potential to alter the calorimeter's energy equivalence or introduce additional heat, resulting in inaccuracies in the determined calorific value (Mesroghli, Jorjani, & Chelgani, 2009).

The requirements for the experiment necessitate a starting oxygen pressure ranging from 300 to 600 psi and an ending temperature within the range of 20 to 35°C (68 to 95°F). The resulting products are ashes, water, carbon monoxide, sulfur dioxide, and nitrogen. Upon determination of the gross caloric value, the net calorific value (NCV), which represents the net warmth of burning, is subsequently computed by subtracting 1,030 Btu/lb (2.4 × 103 kJ/kg) via the GCV [at 20°C (68°F)] to account for the thermal energy of condensation of the water. The deduction of the heat for [1,055 Btu/lb (2.45 × 103 kJ/kg)] is not equivalent to the actual value. The computation aims to convert the data via a gross amount at a constant size to a net quantity during constant pressure (Matin & Chelgani, 2016). The following delineates the distinctions across the GCV along with the NCV.

$$\text{NCV (Btu/lb)} = \text{GCV} - \frac{1030 \times \%\ \text{total hydrogen} \times 9}{100}$$

The GCV is the documented amount for both forms of measurement. The net calorie content can be derived by deducting an appropriate amount from the total calorific value [measured at 20°C (68°F)], which accounts for the moisture content present in the coal before and during combustion. This deduction amounts to 1,030 Btu/lb, 572 cal/g, or 2.395 MJ/g. In this case, the deduction does not correspond to the latent energy for the vaporization process in water, which is 1,055 Btu/lb (2.4 MJ/g) beginning at 20°C (68°F). This is because the deductions are calculated to convert from the total amount at a constant volume to the net amount at a steady pressure, and the correction factor, according to the given circumstances, is 1,030 Btu/lb (2.395 MJ/g) (Chelgani, 2021).

A contemporary technique for measuring coal's calorie content involves utilizing a droperidol and adiabatic bomb, the calorie meter. The former is a calorimeter that features a jacket with a consistent and unchanging temperature.

In the context of adiabatic calorimetry, it is imperative to ensure that the jacket's temperature is appropriately regulated to correspond with that of the thermometer's vessel throughout the temperature increase. It is imperative to maintain a high degree of similarity between the two temperatures during the phase of swift escalation. The droperidol calorimeter may necessitate adjusting radiation to account for temperature rise. If there is sufficient evidence of uncompleted combustion, it is recommended to reject a person's test in both scenarios. Moreover, the monthly inspection of heat capacity may prove insufficient despite being a mandatory requirement. It is advisable for laboratories that conduct a high volume of daily tests to increase the frequency of their heat capacity value assessments. The optimal frequency for conducting heat capacity checks must be established to minimize the impact of an unnoticed shift in warmth and the capacity values on the total amount of tests (Shirazi, Börtin, Eklund, & Lindqvist, 1995).

It is imperative to exercise prudence throughout the collection of samples for calorific value tests, as oxidation of fossil fuels sampling may lead to a decrease in calorific value. Coal specimens of lower rank, such as the lignite and sub-bituminous, can experience more significant burning effects than those of higher-rank fossil fuels samples. It is advisable to prevent any superfluous contact of the specimens with the atmosphere during the sampling process or any postponement in the analysis. Due to the testing technique's inherent characteristics, the calorimeter's components must be scrutinized after each utilization. Furthermore, in the event of any accidental impact or fall, it is recommended that a comprehensive examination of all the constituent parts be conducted before any further usage. The consistent breaks from both measurement techniques exhibit proximity. However, the presence of humans in the equation may be mitigated by using a calorimeter equipped with a computer programmer and a digital temperature gauge (Mahapatra, 2016).

If a coal sample lacks a measured calorie content, it is feasible to approximate the CV through diverse formulas. Among these formulas, the most widely employed are:

1. The Dulong formula is a scientific equation used in the field of thermodynamics:

$$CV = 144.4(\%C) + 610.2(\%H) - 65.9(\%O) - 0.39(\%O)^2$$

2. The formula known as the Dulong-Berthelot calculation is a topic of interest in academic circles:

$$CV = 81{,}370 + 345\left[\frac{\%H - (\%O + \%N - 1)}{8}\right] + 22.2(\%S)$$

The nitrogen, oxygen, carbon, and sulfur (organic) contents of coal are denoted by %C, %H, %N, %O, along with %S, respectively. These values are determined based on a dry, ash-free basis. In both instances, the computed values exhibit a high degree of concurrence with the empirical calorimetric values (Feng, Zhang, Zhang, & Wen, 2015).

In certain cases, the unavailability of an adiabatic bomb thermometer or the insufficiency of the sample size may impede its precise application. To address these issues, it has been demonstrated that thermal differentiation (DTA) is a viable method for determining the energy content of coal. The data acquired through the differential thermal analysis (DTA) technique exhibit a high degree of concurrence with the data obtained via the utilization of the bomb calorimeter (Büyükkanber, Haykiri-Acma, & Yaman, 2023).

6.2.2. Thermal Conductivity

The term "thermal conductivity" refers to the speed at which heat is transferred via conduction over a given area, thickness, and temperature difference:

$$Q = k\frac{A(t_2 - t_1)}{d}$$

The expression for heat, denoted as Q, can be represented in units of kcal/s cm °C or Btu/ft-hr-°F (where 1 Btu/ft-hr-°F is equivalent to 1.7 J/s · m · K). The variable k represents thermal conductivity, A represents area and $t_2 - t_1$ represents the variation in temperature over the distance *d* (Ramazanova, Abdulagatov, & Ranjith, 2018).

As Speight (1994) and other cited sources discussed, joining bedding layers in coal can introduce complexities that hinder determining a definitive thermal insulation value for a given coal. The task may prove to be exceedingly challenging, if not altogether unfeasible. However, deducing specific inferences from the available data has been feasible. The term monolithic coal is characterized as a substance with moderate heat conductivity. Specifically, anthracite coal exhibits thermal conductivity ranging from five to nine $\times 10^{-4}$ kcal/s·cm·°C, whereas monolithic bitumen coal displays a thermal conductivity within a spectrum of four to seven $\times 10^{-4}$ kcal/s·cm·°C. Moreover, it is worth noting that the pulverized coal's heating capacity is comparatively lower than that of its monolithic counterpart. The thermal conductivity coefficients of crushed bituminous coal are within the range of 2.5 with 3.5×10^{-4} kcal/s·cm·°C (Herrin & Deming, 1996).

The relationship between the heating capacity of coal and its apparent density, moisture content, amount of ash, and temperature is generally positive, indicating that an increase in any of these factors is likely to boost thermal conductivity. Furthermore, it has been observed that the heating capacity of coal in the direction parallel to the bedding plane exhibits a greater magnitude than that of the thermal conductivity in the direction parallel to the bedding plane (Rezaei et al., 2000).

Limited data are available regarding water's impact on coal's heating capacity. However, it can be hypothesized that the presence of water in coal may increase its thermal conductivity, given that water's thermal and electrical conductivity is approximately threefold higher than coal's (Zhou, Zhou, Li, & Zhang, 2013).

6.2.2.1. Data Handling and Interpretation

The principal application of plasticity data about materials is to evaluate the coking characteristics of coal. The utilization of maximum fluidity values is a common practice in this regard. However, coal's plastic spectrum was also employed as a criterion for the amalgamation of coals for carbonization. The temperature range at which a material exhibits plastic behavior lies between its softening and solidification temperatures. It is advisable to exercise caution when interpreting plastic assets data. Applying this data is likely to be more advantageous in evaluating variations in the coking properties of the low fluid, less coking-prone coal types, as opposed to assessing discrepancies in the coking traits of high flexible, more coking-prone coal types (Wildemuth & Williams, 1985).

6.2.3. Thermal Conductivity

Coal exhibits moderate thermal conductivity, typically falling within the range of 3×10^{-4} to 9×10^{-4} cal/s · centimeters · °C. The electrical conductivity (λ) of coal is correlated with the heating ability (Cp) and thermal dispersion (α), as per established academic literature.

$$\alpha = \frac{\lambda}{\rho C_p}$$

The symbol 'ρ' denotes the density of the material. Thermal conductivity refers to the heat transfer rate through the conduction of each unit area and unit thickness about a single temperature gradient. Thermal diffusivity refers to quantifying the velocity at which a heat wave propagates (Feng et al., 2015).

The heat transfer capacity and diffusivity of fossil fuels can be determined through steady-state regular warming unsteady-state techniques. The chemical bar technique and thermal imaging have been superseded by the brief hot-wire/line source of heat and temporary hot plate methods, which enable swift and autonomous determinations of α and λ. Indeed, these techniques provide the added benefit of assessing these characteristics for solid samples, coal gathers, and granules in conditions comparable to those observed in coal usage systems (Asadi, Shafigh, Hassan, & Mahyuddin, 2018).

The heat conductivity/diffusivity measurements for coal rank do not exhibit a systematic connection owing to the significant contributions from minerals, with pyrite particularly noteworthy. Conversely, there has been an observed moderate correlation with unadjusted coal densities, as reported by Speight (1994) and the sources referenced therein. The properties above demonstrate anisotropic behavior, whereby the thermal conductivity is maximized for specimens oriented parallel with the mattress's plane (Zorić et al., 2012). The augmentation of obvious density, uncertain matter, and ash, along with mineral matter content, results in an elevation of thermal conductivity. The heating capacity of coal is significantly influenced by its high porosity and the characteristics of the gas, vapor, or water present within the pores, regardless of the case of monolithic samples. The moisture in coal has a comparable impact on its thermal conductivity, as the heat transfer capacity of moist coal is roughly three times greater than that of coal in a dry state. The thermal dispersion of coal remains largely unaltered by moisture, as changes in moisture content do not significantly impact the λ/Cp ratio (Strezov, Lucas, & Strezov, 2000).

There is yet to be a comprehensive model that can be applied to predict heat conductivity/diffusivity values. However, it has been suggested that additive donations from the substances present in coal can be assumed, leading to the proposal of the next correlation:

$$\lambda^{-1}(\mathrm{mK/W}) = \frac{\mathrm{C}}{1.47} + \frac{\mathrm{H}}{0.0118}\left(\frac{273}{T}\right)^{0.5}$$

The equation presented in the text relates to the weight fractions of carbon (C) and hydrogen (H), as well as the temperature in absolute terms in Kelvin (T). It has been observed that this equation provides satisfactory forecasts for monolithic bitumen coal (Du, Yang, Ge, Wang, & Liu, 2021).

The heating capacity of fragmented coal is comparatively inferior to that of solid coal. The model above, which pertains to two-phase systems comprising a non-continuous phase composed of coal particles as well as an ongoing gas phase, is generally consistent with existing literature and forecasts that the heat conductivity of solid bituminous coals is over two times greater than that corresponding to their aggregates in it (Yang, Park, Kim, & Yoo, 2021).

6.3. SUMMARY

The specific thermal and mechanical properties of coal can vary significantly depending on its type, rank, and geographical location. The heating value of coal refers to the amount of heat energy released when it is burned, and its value can vary depending on composition. Coal contains volatile matter which is released as gases and vapors during combustion. Coal's mechanical properties, such as hardness, can vary depending on its rank or maturity level. Higher-rank coals, such as anthracite are generally harder and more brittle. Some types of coal such as bituminous coal possess coking properties.

REFERENCES

1. Ai, W., Xue, B., Wei, C., Dou, K., & Miao, S., (2018). Mechanical and thermal properties of coal gasification fine slag reinforced low density polyethylene composites. *Journal of Applied Polymer Science, 135*(17), 46203.
2. Aida, T., & Okamoto, T., (1960). Determination of the workability of a coal face underground by measuring the friability of coal. *Memoirs of the Faculty of Engineering, Kyoto University, 22*(4), 387–400.
3. Akram, M., Tan, C., Garwood, R., & Thai, S., (2015). Vinasse – A potential biofuel – cofiring with coal in a fluidized bed combustor. *Fuel, 158*, 1006–1015.
4. Altun, N. E., Hicyilmaz, C., & Kök, M. V., (2001). Effect of different binders on the combustion properties of lignite part I. Effect on thermal properties. *Journal of Thermal Analysis and Calorimetry, 65*, 787–795.
5. Anderson, K., Scales, P., & Ralston, J., (1986). Dustiness and wettability of very fine Australian coal particles. *AIMM Bull. Proc. (Australia), 291*(3).
6. Asadi, I., Shafigh, P., Hassan, Z. F. B. A., & Mahyuddin, N. B., (2018). Thermal conductivity of concrete – A review. *Journal of Building Engineering, 20*, 81–93.
7. Auvil, H., & Davis, J., (1938). *Determination of the Swelling Properties of Coal During the Coking Process* (Vol. 3403): US Department of the Interior, Bureau of Mines.
8. Baldwin, G., & Vynne, J., (1995). *The Application and Economic Benefits of Blasthole Drill Monitors in Coal Mines.*
9. Bhatt, A., Priyadarshini, S., Mohanakrishnan, A. A., Abri, A., Sattler, M., & Techapaphawit, S., (2019). Physical, chemical, and geotechnical properties of coal fly ash: A global review. *Case Studies in Construction Materials, 11*, e00263.
10. Bilgen, S., & Kaygusuz, K., (2008). The calculation of the chemical exergies of coal-based fuels by using the higher heating values. *Applied Energy, 85*(8), 776–785.
11. Bilgen, S., Keleş, S., & Kaygusuz, K., (2012). Calculation of higher and lower heating values and chemical exergy values of liquid products obtained from pyrolysis of hazelnut cupulae. *Energy, 41*(1), 380–385.
12. Brewer, R. E., (1942). *Plastic and Swelling Properties of Bituminous Coking Coals* (Vol. 445). US Department of Interior.
13. Büyükkanber, K., Haykiri-Acma, H., & Yaman, S., (2023). Calorific value prediction of coal and its optimization by machine learning based

on limited samples in a wide range. *Energy, 277*, 127666.

14. Chelgani, S. C., (2021). Estimation of gross calorific value based on coal analysis using an explainable artificial intelligence. *Machine Learning with Applications, 6*, 100116.
15. Czapliński, A., & Hołda, S., (1982). Changes in mechanical properties of coal due to sorption of carbon dioxide vapor. *Fuel, 61*(12), 1281–1282.
16. Davis, J. D., (1931). The plastometer: A new instrument for measuring plastic properties of coal. *Industrial and Engineering Chemistry Analytical Edition, 3*(1), 43–45.
17. Davis, J. D., (1943). *Carbonizing Properties and Petrographic Composition of Thick Freeport-Bed Coal from Harmar Mine, Harmarville, Allegheny County, Pa: And the Effect of Blending This Coal with Pocahontas No. 3-and 4-bed Coals* (Vol. 655). US Government Printing Office.
18. Davis, J. D., Jung, F., Juettner, B., & Wallace, D., (1933). Plastic properties of cooking coals. *Industrial and Engineering Chemistry, 25*(11), 1269–1274.
19. Deng, J., Li, Q.-W., Xiao, Y., & Shu, C.-M., (2017). Experimental study on the thermal properties of coal during pyrolysis, oxidation, and re-oxidation. *Applied Thermal Engineering, 110*, 1137–1152.
20. Deniz, V., & Umucu, Y., (2013). Interrelationships between the Bond grindability with physicomechanical and chemical properties of coals. *Energy Sources, Part A: Recovery, Utilization, and Environmental Effects, 35*(2), 144–151.
21. Du, Y., Yang, W., Ge, Y., Wang, S., & Liu, P., (2021). Thermal conductivity of cement paste containing waste glass powder, metakaolin, and limestone filler as supplementary cementitious material. *Journal of Cleaner Production, 287*, 125018.
22. Dumansky, A., Tairova, L., Gorlach, I., & Alimov, M., (2011). A design-experiment study of nonlinear properties of coal-plastic. *Journal of Machinery Manufacture and Reliability, 40*(5), 483.
23. Falcon, L., & Falcon, R., (1987). The petrographic composition of Southern African coals in relation to friability, hardness, and abrasive indices. *Journal of the Southern African Institute of Mining and Metallurgy, 87*(10), 323–336.
24. Feng, J., Zhang, R., Gong, L., Li, Y., Cao, W., & Cheng, X., (2015). Development of porous fly ash-based geopolymer with low thermal conductivity. *Materials and Design (1980–2015), 65*, 529–533.
25. Feng, Q., Zhang, J., Zhang, X., & Wen, S., (2015). Proximate analysis-

based prediction of gross calorific value of coals: A comparison of support vector machine, alternating conditional expectation and artificial neural network. *Fuel Processing Technology, 129*, 120–129.

26. Fernández, A., Barriocanal, C., Díez, M., & Alvarez, R., (2009). Influence of additives of various origins on thermoplastic properties of coal. *Fuel, 88*(12), 2365–2372.
27. Gautam, S., Yadav, R., & Suresh, N., (2015). Evaluation of the characteristics of as-received and washed low grade Indian coals for their industrial applications. *Energy Sources, Part A: Recovery, Utilization, and Environmental Effects, 37*(19), 2138–2149.
28. Gil, M., García, R., Pevida, C., & Rubiera, F., (2015). Grindability and combustion behavior of coal and torrefied biomass blends. *Bioresource Technology, 191*, 205–212.
29. Greenhalgh, S., & Emerson, D., (1986). Elastic properties of coal measure rocks from the Sydney Basin, New South Wales. *Exploration Geophysics, 17*(3), 157–163.
30. Guo, L., Zhai, M., Wang, Z., Zhang, Y., & Dong, P., (2018). Comprehensive coal quality index for evaluation of coal agglomeration characteristics. *Fuel, 231*, 379–386.
31. Hamelmann, F., & Schmidt, E., (2003). Methods of estimating the dustiness of industrial powders – A review. *KONA Powder and Particle Journal, 21*, 7–18.
32. Hanrot, F., Ablitzer, D., Houzelot, J., & Dirand, M., (1994). Experimental measurement of the true specific heat capacity of coal and semicoke during carbonization. *Fuel, 73*(2), 305–309.
33. Hauserman, W., & Tye, C., (1984). *Mechanical and Thermal Friability of Coals and Modes of Water Bonding: North Dakota Univ., Grand Forks (USA)*. Energy Research Center.
34. Herrin, J. M., & Deming, D., (1996). Thermal conductivity of US coals. *Journal of Geophysical Research: Solid Earth, 101*(B11), 25381–25386.
35. Hertzog, E. S., & Cudworth, J. R., (1938). *Friability of Alabama Coals* (Vol. 3384). United States Bureau of Mines.
36. Honda, H., Sanada, Y., & Furuta, T., (1966). Mechanical and thermal properties of heat-treated coals. *Carbon, 3*(4), 421-IN415.
37. Jones, A. H., Bell, G. J., & Schraufnagel, R. A., (1988). *A Review of the Physical and Mechanical Properties of Coal with Implications for Coal-Bed Methane Well Completion and Production.*
38. Kalaw, M. E. L., Culaba, A. B., Nguyen, H. T., Nguyen, K., Hinode, H.,

Kurniawan, W., Promentilla, M. A. B., (2015). Mechanical and thermal properties of geopolymers from mixtures of coal ash and rice hull ash using water glass solution as activator. *ASEAN Journal of Chemical Engineering, 15*(2), 51–61.

39. Klawitter, M., Esterle, J., & Collins, S., (2015). A study of hardness and fracture propagation in coal. *International Journal of Rock Mechanics and Mining Sciences, 76*, 237–242.
40. Laubach, S., Marrett, R., Olson, J., & Scott, A., (1998). Characteristics and origins of coal cleat: A review. *International Journal of Coal Geology, 35*(1–4), 175–207.
41. Li, M., Mao, X., Cao, L., Pu, H., Mao, R., & Lu, A., (2016). Effects of thermal treatment on the dynamic mechanical properties of coal measures sandstone. *Rock Mechanics and Rock Engineering, 49*, 3525–3539.
42. Lu, S., Zhang, Y., Sa, Z., & Si, S., (2019). Evaluation of the effect of adsorbed gas and free gas on mechanical properties of coal. *Environmental Earth Sciences, 78*, 1–15.
43. Lyons, B., & Wisman, P. S., (2004). Coal bed methane play and prospect evaluations using GeoGraphix software. *ASEG Extended Abstracts, 2004*(1), 1–5.
44. Mahapatra, D., (2016). A review on steam coal analysis-calorific value. *American International Journal of Research in Science, Technology, Engineering, and Mathematics, 14*(1), 57–68.
45. Majumder, A., Jain, R., Banerjee, P., & Barnwal, J., (2008). Development of a new proximate analysis-based correlation to predict calorific value of coal. *Fuel, 87*(13–14), 3077–3081.
46. Maloney, D. J., Sampath, R., & Zondlo, J. W., (1999). Heat capacity and thermal conductivity considerations for coal particles during the early stages of rapid heating. *Combustion and Flame, 116*(1–2), 94–104.
47. Mason, D. M., & Gandhi, K. N., (1983). Formulas for calculating the calorific value of coal and coal chars: Development, tests, and uses. *Fuel Processing Technology, 7*(1), 11–22.
48. Matin, S., & Chelgani, S. C., (2016). Estimation of coal gross calorific value based on various analyses by random forest method. *Fuel, 177*, 274–278.
49. Menéndez, J., Pis, J., Alvarez, R., Barriocanal, C., Fuente, E., & Díez, M., (1996). Characterization of petroleum coke as an additive in metallurgical coke making. Modification of thermoplastic properties of coal. *Energy and Fuels, 10*(6), 1262–1268.

50. Mesroghli, S., Jorjani, E., & Chelgani, S. C., (2009). Estimation of gross calorific value based on coal analysis using regression and artificial neural networks. *International Journal of Coal Geology, 79*(1–2), 49–54.

51. Mikula, R., & Parsons, I., (1991). Coal dustiness: Characterization and control. *Coal Preparation, 9*(3–4), 199–212.

52. Miroshnichenko, D., Desna, N., Koval, V., & Fatenko, S., (2019). Hardgrove grindability of coal. Part 1. Correlations with composition, structure, and properties. *Coke and Chemistry, 62*, 1–4.

53. Mostaghimi, P., Armstrong, R. T., Gerami, A., Hu, Y., Jing, Y., Kamali, F., & Ramandi, H. L., (2017). Cleat-scale characterization of coal: An overview. *Journal of Natural Gas Science and Engineering, 39*, 143–160.

54. Mraw, S., & Naas-O'Rourke, D., (1979). Water in coal pores: Low-temperature heat capacity behavior of the moisture in Wyodak coal. *Science, 205*(4409), 901–902.

55. Nguyen, H., Bui, H.-B., & Bui, X.-N., (2021). Rapid determination of gross calorific value of coal using artificial neural network and particle swarm optimization. *Natural Resources Research, 30*, 621–638.

56. Patel, S. U., Kumar, B. J., Badhe, Y. P., Sharma, B., Saha, S., Biswas, S., & Kulkarni, B. D., (2007). Estimation of gross calorific value of coals using artificial neural networks. *Fuel, 86*(3), 334–344.

57. Pierron, E. D., Rees, O., & Clark, G., (1959). *Plastic Properties of Coal*. State of Illinois, Department of Registration and Education, Division of the State Geological Survey.

58. Ramazanova, A. E., Abdulagatov, I. M., & Ranjith, P. G., (2018). Temperature effect on the thermal conductivity of black coal. *Journal of Chemical and Engineering Data, 63*(5), 1534–1545.

59. Ren, S.-J., Wang, C.-P., Xiao, Y., Deng, J., Tian, Y., Song, J.-J., & Sun, G.-F., (2020). Thermal properties of coal during low temperature oxidation using a grey correlation method. *Fuel, 260*, 116287.

60. Reynolds, D. A., Davis, J. D., Brewer, R. E., Ode, W. H., Wolfson, D. E., & Birge, G., (1946). *Carbonizing Properties of Western Coals* (Vol. 692). US Government Printing Office.

61. Rezaei, H., Gupta, R., Bryant, G., Hart, J., Liu, G., Bailey, C., & Endo, Y., (2000). Thermal conductivity of coal ash and slags and models used. *Fuel, 79*(13), 1697–1710.

62. Sengupta, A. N., (2002). An assessment of grindability index of coal. *Fuel Processing Technology, 76*(1), 1–10.

63. Shirazi, A. R., Börtin, O., Eklund, L., & Lindqvist, O., (1995). The impact

of mineral matter in coal on its combustion, and a new approach to the determination of the calorific value of coal. *Fuel, 74*(2), 247–251.

64. Shkoller, M., Kazimirov, S., Temlyantsev, M., & Basegskiy, A., (2015). Conditioning of coal-enrichment waste with high moisture and ash content. *Coke and Chemistry, 58*, 482–486.
65. Song, Z., & Zhang, J., (2021). The effect of confining pressure on mechanical properties in coal and rock: review and new insights. *Arabian Journal of Geosciences, 14*, 1–22.
66. Strezov, V., Lucas, J. A., & Strezov, L., (2000). Quantifying the heats of coal devolatilization. *Metallurgical and Materials Transactions B, 31*, 1125–1131.
67. Szwilski, A., (1985). Relation between the structural and physical properties of coal. *Mining Science and Technology, 2*(3), 181–189.
68. Tang, F., Wang, L., Lu, Y., & Yang, X., (2015). Thermophysical properties of coal measure strata under high temperature. *Environmental Earth Sciences, 73*, 6009–6018.
69. Tiryaki, B., (2005). Practical assessment of the grindability of coal using its hardness characteristics. *Rock Mechanics and Rock Engineering, 38*(2), 145.
70. Tomeczek, J., & Palugniok, H., (1996). Specific heat capacity and enthalpy of coal pyrolysis at elevated temperatures. *Fuel, 75*(9), 1089–1093.
71. Tyler, R. J., (1980). Flash pyrolysis of coals. Devolatilization of bituminous coals in a small fluidized-bed reactor. *Fuel, 59*(4), 218–226.
72. Van Krevelen, D. W., (1993). *Coal: Typology-Physics-Chemistry-Constitution. 1(2),* 35-70.
73. Wen, H., Lu, J.-h., Xiao, Y., & Deng, J., (2015). Temperature dependence of thermal conductivity, diffusion, and specific heat capacity for coal and rocks from coalfield. *Thermochimica Acta, 619*, 41–47.
74. Wen, X., Jian, S., & Wang, J., (2017). Prediction models of calorific value of coal based on wavelet neural networks. *Fuel, 199*, 512–522.
75. Wildemuth, C., & Williams, M., (1985). A new interpretation of viscosity and yield stress in dense slurries: coal and other irregular particles. *Rheologica Acta, 24*(1), 75–91.
76. Xie, H., Zhao, X., Liu, J., Zhang, R., & Xue, D., (2012). Influence of different mining layouts on the mechanical properties of coal. *International Journal of Mining Science and Technology, 22*(6), 749–755.
77. Yakimov, M., Gan, T., Mazumder, S., & Croon, M., (2014). Locating and quantifying coal cleats with electrical and acoustic imagers for CSG

development. *The APPEA Journal, 54*(2), 538–538.

78. Yancey, H. F., & Zane, R., (1933). *Comparison of Methods for Determining Friability of Coal* (Vol. 3215). Department of Commerce, US Bureau of Mines.
79. Yang, I.-H., Park, J., Kim, K.-C., & Yoo, S.-W., (2021). A comparative study on the thermal conductivity of concrete with coal bottom ash under different drying conditions. *Advances in Civil Engineering, 2021*, 1–12.
80. Yaralı, O., (2017). Investigation into relationships between Cerchar hardness index and some mechanical properties of coal measure rocks. *Geotechnical and Geological Engineering, 35*, 1605–1614.
81. Zhang, H., Liu, J., & Elsworth, D., (2008). How sorption-induced matrix deformation affects gas flow in coal seams: A new FE model. *International Journal of Rock Mechanics and Mining Sciences, 45*(8), 1226–1236.
82. Zhou, H., Zhou, B., Li, L., & Zhang, H., (2013). Experimental measurement of the effective thermal conductivity of ash deposit for high sodium coal (Zhun Dong coal) in a 300 KW test furnace. *Energy and Fuels, 27*(11), 7008–7022.
83. Zhu, W., Liu, J., Sheng, J., & Elsworth, D., (2007). Analysis of coupled gas flow and deformation process with desorption and Klinkenberg effects in coal seams. *International Journal of Rock Mechanics and Mining Sciences, 44*(7), 971–980.
84. Zorić, D., Lazar, D., Rudić, O., Radeka, M., Ranogajec, J., & Hiršenberger, H., (2012). Thermal conductivity of lightweight aggregate based on coal fly ash. *Journal of Thermal Analysis and Calorimetry, 110*(1), 489–495.

CHAPTER

7 Chemical Reactivity of Coal

Contents

INTRODUCTION

Coal is utilized in various industrial processes, including combustion, carbonization, liquefaction, and gasification (Crelling et al., 1988). Coal-derived products serve as precursors for various materials and are utilized in the chemical manufacturing process. It is imperative to conduct a thorough characterization of coal, whether it is used as a singular entity or in combination with other coals. This is done to ascertain its properties, evaluate its quality, and anticipate its technological performance.

Moreover, comprehending the chemical properties of coal is a crucial aspect in predicting the effective application of coal, such as its utilization as a chemical source for transformation and utilization procedures.

Since the inception of coal science, a significant body of research has been conducted on the diverse facets of coal chemistry. In addition, the inclination towards utilizing coal in order to generate gaseous and liquid fuels has resulted in a significant surge in research endeavors pertaining to the chemical processes of coal. The lack of a cohesive and shared objective in the past has resulted in a scattered approach to studying the chemical properties of coal (Chan et al., 1999).

The reactive nature of coal cannot be accurately anticipated through chemical or physical analyses due to its intricate composition of macromolecules and a physical amalgamation of inorganic and organic constituents. There is a body of evidence to suggest that the reactions occurring within organic materials may be impacted by the presence of minerals such as clay and pyrites that are dispersed throughout the organic coal matrices (Zhang et al., 2015). An expedient approach to comprehending reactions in each system is to scrutinize the reactive functional groups.

Despite the fact that coal chemistry is frequently depicted by one or more basic chemical equations, it encompasses considerably more than what these equations suggest, including diffusional limitatia As an illustration, the process of coal combustion is typically depicted as the chemical reaction between the carbon present in oxygen and coal. The stoichiometry of this reaction is consistently established and maintained (Shao et al., 2023).

$$C_{coal} + O_2 \text{®} 2CO$$

$$C_{coal} + O_2 \text{®} CO_2$$

$$C_{coal} + O2 \text{®} CO_2$$

$$C_{coal} + H_2O \text{®} CO + H_2$$

Thus,

$C_{coal} + O_2$ ® $2CO$

$2CO + O_2$ ® $2CO_2$

Reactions involving hydrogen, such as the one represented by the equation $2H_{coal} + O_2 \rightarrow H_2$, are regarded as being of subordinate significance. It is imperative to acknowledge the intricate nature of coal and the potential impact of heteroatoms, namely nitrogen, oxygen, and sulfur, on the combustion process (Arenillas et al., 2023). This influence can lead to significant environmental issues that warrant attention.

The conversion of sulfur and nitrogen into their corresponding oxides, as depicted by simplified chemical equations, is a significant concern during the process of combustion (Molina & Mondragon, 1998).

$S_{coal} + O_2$ ® SO_2

$2SO_2 + O_2$ ® $2SO_3$

$N_{coal} + O_2$ ® $2NO$

$2NO + O_2$ ® $2NO_2$

$N_{coal} + O_2$ ® NO

Furthermore, the sulfur dioxide that is released into the atmosphere can either be deposited in the nearby vicinity or undergo additional transformation into sulfurous and sulfuric acids through its reaction with atmospheric moisture (Li et al., 2021).

$SO_2 + H_2O$ ® H_2SO_3

$2SO_2 + O_2$ ® $2SO_3$

$SO_3 + H_2O$ ® H_2SO_4

Thus,

$2SO_2 + O_2 + 2H_2O$ ® $2H_2SO_4$

The chemical properties of nitrogen species present in coal hold significant importance. Nitrogen oxides are known to contribute to the creation and production of acid rain, similar to the way sulfur oxides produce acids, resulting in the formation of nitrous and nitric acids (Fung & Kim, 1983).

$NO + H_2O$ ® H_2NO_2

$2NO + O_2$ ® $2NO_2$

$NO_2 + H_2O$ ® HNO_3

Thus,

$2NO + O_2 + H_2O$ ® $2HNO_3$

The chemical processes that occur during coal combustion within combustors are considerably intricate and multifaceted, surpassing the level of complexity that can be adequately conveyed through the utilization of these rudimentary equations (Osório et al., 2006). However, these serve as an initial starting point. Although there have been advancements in comprehending the intricate molecular composition of coal, a definitive structure for coal has yet to be established with complete certainty. Consequently, explanations of the chemical processes of coal are frequently oversimplified and misunderstood, resulting in only cursory acknowledgement at best. The significance of coal chemistry in the context of coal utilization cannot be overstated. Efficient methods of conversion can be achieved through the manipulation of chemical factors. Comprehending the chemistry involved necessitates a comprehensive grasp of coal's behavior in diverse reaction systems.

It is noteworthy to mention that several chemical reactions have been utilized on coal to identify or approximate the structural and functional categories present in coal. This information is crucial to consider when studying the chemistry of coal, as any resultant structures must align with the chemical and physical properties of the coal.

Hence, the current chapter endeavors to provide a relevant overview of the chemical properties of coal and elucidate the potential applications of this knowledge in delineating the performance of coal in diverse conversion methodologies. Incorporating the entirety of coal reactions within this chapter would have been deemed unsuitable. It is imperative that numerous reactions are incorporated in other sections of this document, and solely the reactions that are relevant to the significance of this particular chapter are presented herein.

7.1. REACTIONS WITH OXYGEN

The process of oxidation is multifaceted, involving concurrent yet interdependent reaction mechanisms. While it is widely accepted that there are a minimum of four processes of this nature, the precise quantity, characteristics, and kinetics of these processes remain somewhat ambiguous. The process of aerial oxidation of coal is primarily a chemical reaction, however, it is subject to various external factors such as humidity, temperature, catalytic impacts of water, and mineral matter components, in addition to its inherent chemical composition. Moreover, the impact of both surface and physical properties remains inadequately comprehended.

The process of coal oxidation in the presence of air is a crucial aspect that necessitates consideration when evaluating coal quality, as it significantly impacts the technological characteristics of coal (Zhao et al., 2021). The oxidation of coal may occur due to the effects of weathering processes during transportation and handling, or when the coal is stored under varying environmental conditions. This can lead to the degradation of the inorganic as well as organic components of coal, ultimately resulting in a decline in its properties.

An alternative outcome resulting from coal oxidation is the occurrence of spontaneous combustion, wherein the intrinsic oxidation process generates sufficient heat to initiate smoldering and eventual combustion of the coal, without any external heat input (Kogelschatz et al., 1988). The susceptibility to oxidation is primarily influenced by the coal's rank, potentially in combination with the maceral composition and mineral matter content. Coal with low rank exhibits a heightened susceptibility to spontaneous combustion. To mitigate the constant threat of unprompted ignition and combustion, it may be necessary to regulate variables such as the availability of air to coal stockpiles.

The process of coal oxidation has garnered significant attention as a potential method for altering the coal's chemical and physical characteristics. The oxidation treatment has the potential to mitigate or eliminate the coal caking phenomenon during carbonization (Yeow et al., 2018). The oxidation level holds significant importance, with approximately 1% of oxygen getting absorbed onto or combined with the coal. Chemical examination or trial carbonization can be utilized to estimate it. However, in order to monitor oxidation plant and promptly detect any deviation from the appropriate operating conditions, a faster method of indication is necessary.

The oxidation reaction of coal can be influenced by factors beyond the organic matter and oxygen due to its heterogeneous nature (Schwitzgebel et al., 1995). The aforementioned factors, namely moisture, thermal chemistry, minerals, particle size, and surface area, are taken into account. Nonetheless, there is still a lack of comprehensive comprehension regarding the parameters that influence the process of natural oxidation. Additionally, there are unsettled discrepancies regarding the actual characteristics of the structural modifications that occur in organic matter as a result of oxidation.

The impact of oxidation on coal behavior during combustion, beneficiation, and processing, particularly carbonization, is predominantly negative.

The investigation of the reaction between coal and atmospheric oxygen has been a topic of extensive research. The investigation of weathering phenomena, the integration of functional categories, and the mechanisms underlying the unforeseen ignition of coal are among the numerous areas of research that have garnered significant attention (Ritchie, 1981).

Conversely, the oxidation of coal that has undergone weathering is considerably more challenging, and any subsequent oxidation transpires with significant difficulty, if it occurs at all. Apart from its impact on the carbonizing characteristics of coal, atmospheric oxidation is believed to diminish the quantity of coal-soluble substances (Good et al., 1970). However, certain coal types have been observed to exhibit a rise in extractable matter, specifically humic acids, following weathering. This phenomenon enhances coal's flotation response, thereby conferring an advantage for coal cleaning procedures. The complexities of coal chemistry are further compounded by seemingly paradoxical observations of this nature, which frequently entail challenging or unfeasible direct comparisons of experimental conditions. The perceived divergent characteristics of the information are contingent upon the delineation of retrievable substance (Zhang et al., 2018). It can be postulated that the presence of organic nonpolar substance, which is susceptible to extraction by organic solvents, may exhibit a reduction upon undergoing oxidation. Conversely, it is reasonable to anticipate a reduction in polar substances, such as humic acids, resulting from coal oxidation. The issue at hand could potentially be attributed to a matter of defining terms.

A comparison has been made between the low-temperature oxidation and combustion of coal with respect to the gases produced (carbon monoxide, CO; carbon dioxide, CO_2; and water, H_2O), which are identical in both processes (Jörissen, 2006). However, a significant disparity arises in the reaction rate. Additionally, it is noteworthy that during oxidation, up to 50% of the oxygen utilized may persist in the coal and may even result in an increase in weight. The retention of oxygen in coal has been attributed to the formation of a coal-oxygen complex, which subsequently undergoes decomposition to yield carbon monoxide, carbon dioxide, and water.

7.1.1. Chemical and Physicochemical Aspects

It is postulated that coal undergoes a reaction with oxygen at standard temperatures through the temporary creation of a coal-oxygen complex, also known as a "oxy-coal" complex. The phenomenon known as oxy-coal is discernible through the microscopic analysis of coal, which reveals the presence of oxidation edges and an increase in reflective material, as depicted in Figure 7.1 (Cybulska et al., 2008).

It is postulated that the process of coal oxidation at low temperatures occurs through three discernible stages: firstly, the adsorption of oxygen onto the coal surface resulting in the formation of an oxygen complex alongside the coal; secondly, the breakdown of the coal-oxygen complex; and finally, the creation of oxy-coal. It is postulated that the coal surface undergoes chemisorption of oxygen, resulting in the formation of a hydroperoxide and/or peroxide (Kuznetsov & Kazansky, 2008). The precise nature of the peroxide formed is currently speculative and may differ depending on the specific site of interaction. The exothermicity of this reaction is estimated to be approximately 100 kcal/mol oxygen.

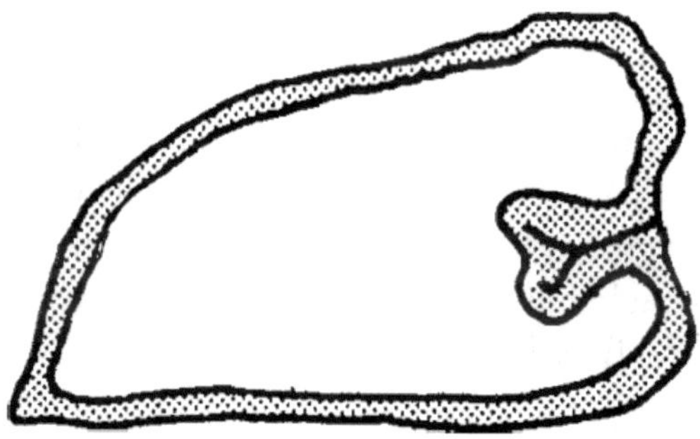

Figure 7.1. *Illustration of a coal particle's oxidation edge (the darkened area).*

Source: James G. Speight, Creative Commons License.

Figure 7.2 depicts the proposed intermediate hydroperoxide that results from the presence of unpaired electron sites within the coal (Nassimbeni, 2003). The subsequent oxidation of coal generates oxygen-containing functional groups such as carboxylic acid, carbonyl, and phenolic hydroxyl. Alkali-soluble compounds, known as humic acids, are produced during the coal substance's breakdown process. These intermediates then undergo further breakdown, yielding molecules with even lower molecular weights.

Notwithstanding oversimplifications, it is important to note that the process of aerial oxidation of coal is highly intricate. During the early phases, a minor increase in weight (attributable to the inclusion of oxygen while retaining all volatile oxidation byproducts) may be observed, succeeded by a conspicuous reduction in the hydrogen and carbon constituents of the resultant substance(s), ultimately converging towards the seemingly "restrictive" thresholds of approximately 60%–65% carbon and 2%–3% hydrogen, contingent on the quality of the initial coal and the oxidation circumstances (primarily temperature) (Jiménez-Escrig & Sánchez-Muniz, 2000).

$$(\text{Coal}){:}\text{H} + \text{O}_2 \rightarrow (\text{coal})\cdot + \text{HO}_2\cdot$$

$$(\text{Coal})\cdot + \text{O}_2 \rightarrow (\text{coal})\text{O}_2\cdot$$

$$(\text{Coal})\text{O}_2\cdot + (\text{coal}){:}\text{H} \rightarrow (\text{coal})\text{OOH} + (\text{coal})\cdot$$

Figure 7.2. *Coal oxidation caused by the production of a momentary hydroperoxide.*

Source: Marcel Dekker, Creative Commons License.

Elevated temperatures, typically at 70°C (160°F), tend to expedite the dehydrogenation process or the reactivity of aromatic systems. However, the product distribution is largely dependent on the thermal endurance of the transient intermediates. At a temperature of 72°C (160°F), the reaction typically does not progress to the extent of significant degradation of the coal substance (Mareschal et al., 2010). Instead, it typically concludes with the creation of diverse functional groups of oxygen within the coal.

Nevertheless, at elevated temperatures (specifically at 70°C [160°F]), the pace of oxidation is significantly amplified and is also contingent on the coal's porosity. Within this spectrum, it is plausible that an inverse relationship exists between the rate of oxidation and either the size of the particle or the degree of rank escalation (Barrera-Díaz et al., 2011). The gaseous products may exhibit a rise in the concentration of carbon dioxide.

Humic acids are formed primarily at temperatures exceeding 70°C. At higher temperatures, ranging from 150°C to 250°C (300°F to 480°F), these acid products are deemed stable. This is noteworthy as petroleum asphaltene constituents, which are complex natural products containing oxygen functions, release oxygen-containing gases, likely from the oxygen-containing functions, at temperatures between 150°C and 300°C (300°F to 570°F) (Lieberman & Vemuri, 2015). While humic materials and petroleum asphaltene constituents may be distantly related, the stability of humic acids at high temperatures is surprising.

Humic acids in their undiluted form are dark, shiny solids that contain approximately 55% to 60% carbon and 2% to 3% hydrogen. The remaining elemental composition is primarily composed of oxygen. The molecular masses of humic acids exhibit significant variability, necessitating the assignment of a range between 600 and 10,000 for these weights (Guy & Hadgraft, 1988). The observed prevalence of varying molecular weights in these substances can be attributed to their inclination to associate, even in solutions of low concentration. Consequently, the determined molecular weights are significantly influenced by both the solvent utilized for the assessment and the techniques employed.

7.1.2. Effect of Moisture

The impact of moisture on the low-temperature oxidation rate of coal is perceptible (Kurose et al., 2001). Consequently, in the event that the coal is adequately desiccated, the pace of oxidation diminishes, albeit additional variables may potentially contribute to this ostensibly uncomplicated observation. Water is a resultant of the oxidation process (Suuberg et al., 1993). If the pace of oxidation seems to escalate over time, it is plausible that the water produced as a reaction byproduct instigates the development of the fleeting peroxide or hydroperoxide. Conversely, excessive desiccation of coal can have negative consequences on its pore structure, leading to a significant decrease in the rate of oxidation due to a reduction in the available surface area.

7.1.3. Effect of Pyrite

The minerals pyrite (FeS_2) and marcasite (FeS_2) are commonly found in coal and are acknowledged as the primary mineral components in numerous coal deposits (Arisoy & Beamish, 2015). Consequently, it is unsurprising that there exists a degree of scholarly attention directed towards examining the impact of these substances (with a particular emphasis on pyrite, given its widespread prevalence) on the process of coal oxidation.

The exothermic nature of the lower temperatures oxidation of pyrite has been observed (Deng et al., 2015).

$$2FeS_2 + 7O_2 + 2H_2O \text{ ® } 2H_2SO_4 + 2FeSO_4 + 62 \text{ kcal/mol}$$

Thus, storing coal where it would be repeatedly wet (by rain) may favor pyrite oxidation.

The oxidation of pyrite may be facilitated by the repeated exposure of coal to moisture, such as rainfall, during storage (Pietrzak et al., 2007).

If coal is preserved by being fully submerged in water, it would prevent the aforementioned scenario. Despite the exothermic nature of pyrite oxidation, it seems that pyrite's role in coal oxidation is insignificant. Nevertheless, some studies have proposed that pyrite oxidation could potentially contribute to the self-ignition of coal.

7.1.4. Spontaneous Ignition

The phenomenon of coal's spontaneous ignition, also known as autogenous heating or spontaneous combustion, has been acknowledged as a potential hazard for a considerable period (Jackson, 1951). In fact, guidelines were established in the early 1900s with the sole objective of reducing the occurrence of self-heating and have undergone revisions since then. The occurrence of unprompted ignition is not exclusive to coal, as it has been documented in various accumulations of organic matter. The phenomenon of coal self-heating, also known as spontaneous combustion, is a naturally transpiring process that arises from the oxidation of coal. The phenomenon of coal self-heating is influenced by a variety of factors, both within and outside of human control (Zabetakis et al., 1954). The controllable factors pertaining to coal-fired power plants encompass the close supervision of coal storage in silos/bunkers, stockpiles, and mills, as well as the management of coal transportation. Uncontrollable variables encompass the inherent properties of coal as well as the surrounding environmental conditions.

The phenomenon of coal self-heating can be attributed to a series of intricate exothermic chemical reactions (Astbury & Hawksworth, 2007). Assuming an uninterrupted air supply and absence of heat dissipation, coal has the propensity to undergo self-heating. The phenomenon of coal self-heating is influenced by a multitude of factors that can be broadly classified into two categories: intrinsic factors pertaining to the coal's properties and extrinsic factors related to the storage conditions and environment. The process of self-heating can lead to the deterioration of coal, thereby altering its chemical and physical properties, which in turn can have a significant impact on the operational efficiency of boilers.

In certain instances, there are concerns regarding the potential for spontaneous combustion during the final preparation stages, such as in silos, bunkers, and mills. The self-heating tendency of coal is influenced by various properties such as the volatile material, coal size of particles, rank, thermal capacity, heat that comes from a reaction, oxygen content, and pyrite content (Freeman & Lefebvre, 1984). The likelihood of coal to undergo self-heating and spontaneous combustion has a tendency to escalate as its rank decreases. It has been observed that lignites and sub-bituminous coals exhibit a greater susceptibility to spontaneous combustion in comparison to bituminous coals and anthracites.

The temperature of coal undergoes a process of self-heating until it attains a plateau, at which point the temperature experiences a temporary stabilization (Griffiths & Scott, 1987). Currently, the thermal energy produced through the process of oxidation is utilized to facilitate the conversion of moisture present in coal into a gaseous state. Upon complete vaporization of moisture, there is a rapid escalation in temperature. Conversely, desiccated matter has the propensity to ignite effortlessly upon absorption of water. It is not advisable to store arid coal in a moist location as this can stimulate self-combustion. Consequently, it is advisable to store wet and dry coal in distinct locations.

The phenomenon of spontaneous combustion is a common occurrence in coal stockpiles, particularly in cases where substantial amounts of coal are stored over prolonged periods of time. Coal types with high susceptibility to self-heating, namely lignite, sub-bituminous coal, and brown coal, are seldom retained for extended periods at power plants (Tadema & Wiejdema, 1970). Self-heating is a phenomenon that is frequently observed. The phenomenon of self-heating in coal is an inherent process that arises due to the oxidation of coal. Unregulated natural oxidation has the potential to result in the discharge of pollutants and the occurrence of self-ignition. If not managed appropriately, the outcomes can lead to severe harm to the power generation infrastructure. The present piece of writing constitutes a contemplation on the comparative duration of storage periods associated with each stage. The occurrence of spontaneous combustion within stockpiles presents notable challenges in terms of safety, environmental impact, economic implications, and handling considerations.

Coal poses various risks from the point of extraction to its ultimate utilization in steam boilers and furnaces (Wen et al., 2017). The following are enumerated some of the attributes of unpremeditated coal fires that can be employed to assess the possibility of coal fires and as principles for reducing the likelihood of a fire. There is a positive correlation between the inherent moisture content and the tendency for heating. A positive correlation exists between the heating tendency and the ash-free Btu, whereby a decrease in the latter leads to an increase in the former. A positive correlation exists between the oxygen content present in coal and its heating tendency.

7.2. REACTIONS WITH OXIDANTS

During the nascent stage of coal science, the primary motivation for the oxidation of coal was the synthesis of chemicals (Beccalli et al., 2007). However, despite economic factors, the fundamental objective of identifying low-molecular-weight products and establishing their correlation with coal structure remained a persistent theme.

Given the current focus on understanding the composition of the coal molecule, it is unsurprising that oxidation techniques have regained prominence in the field of coal science. Some engineers, researchers, and scientists argue that this emphasis on elucidating coal's structure has contributed to this trend. Recently, these techniques have been employed to project product data onto different structural entities. The non-oxygen-based oxidation of coal constitutes a significant area of study within the field of coal science. Various oxidants, including nitric acid, permanganate, dichromate, hydrogen peroxide, performic acid, and hypochlorites, have been utilized in the treatment of coal (Marnett et al., 2003).

The investigation of coal oxidation by different oxidants has been a continuous area of research. However, the renewed attention can be attributed to the notion that coal can be classified as a polyadamantane (polyamantane) system. This classification was derived from the analysis of the resulting products when coal interacts with sodium hypochlorite (Halliwell et al., 1988). However, the purpose of this discussion is not to engage in an analysis of the composition of coal resulting from the application of oxidizing agents. Rather, the aim is to provide a succinct overview of the mechanisms by which coal interacts with different oxidants. The surface characteristics of coal are known to be altered by various oxidizing agents, with permanganate exhibiting a notably high efficacy in this regard. Nevertheless, in broader contexts, it seems that coal reacts with oxidants in a non-selective manner, despite multiple assertions regarding the selectivity of different oxidants. The reported conditions for said investigations exhibit significant variation. Nonetheless, it is widely acknowledged that oxidants are capable of attacking coal at a substantially faster rate than atmospheric oxygen or perhaps pure oxygen.

Furthermore, the multifarious nature of the oxidizing agents makes the process of coal oxidation exceedingly intricate, as the experimental conditions may exhibit considerable variation. The heterogeneous nature of coal's structural composition, which exhibits variations not only across different ranks as well as within the same rank, poses significant challenges in the investigation of coal's oxidative degradation (Veiko et al., 2021). The determination of optimal conditions such as time, temperature, and oxidant-to-coal ratio necessitates the conduction of multiple experiments for each oxidant. In addition, it is imperative to take into account the existence of mineral matter as a fundamental component of coal oxidation, given that the chemical composition of the oxidation process may be altered by the presence of mineral constituents. The facilitation of the oxidation reaction can be observed when coal undergoes pretreatment with hydrochloric acid for the purpose of eliminating mineral matter.

The process of coal oxidation typically results in the formation of a diverse array of products, and it is often insufficient to rely on a single solvent extraction method for the complete extraction of all organic acids (Clerici & Ingallina, 1998). In certain cases, it may be imperative to employ a series of solvents or solvent mixtures with varying polarities or elevated temperatures to achieve successful product isolation. Hence, the meticulous selection of appropriate solvents or solvent blends constitutes a significant determinant in the extraction of the organic substances generated through coal oxidation.

Upon their formation, these acids undergo degradation into lower-molecular-weight entities, including aliphatic and aromatic carboxylic acids that are more readily identifiable (Mishra et al., 2005). The ultimate stage involves the generation of carbon dioxide through the aforementioned acids. Acids that possess polycondensed ring systems are infrequently observed in the outcomes of coal oxidation. However, there have been assertions that the mixed aromatic acids derived from coal oxidation with performic acid comprise a diverse range of phenols and alkyl-substituted phenols (as depicted in Figure 7.3) (Keenan & Sedlak, 2008). Despite not being commonly acknowledged as an oxidation method, the utilization of concentrated sulfuric acid for coal treatment results in the integration of oxygen functionalities (such as phenolic hydroxyl, –OH, and carboxylic groups, $–CO_2H$) into the coal. The integration of sulfonic acid functionalities ($–SO_3H$) is observed, albeit typically to a lesser degree than the integration of phenolic functionalities (Traylor et al., 1995).

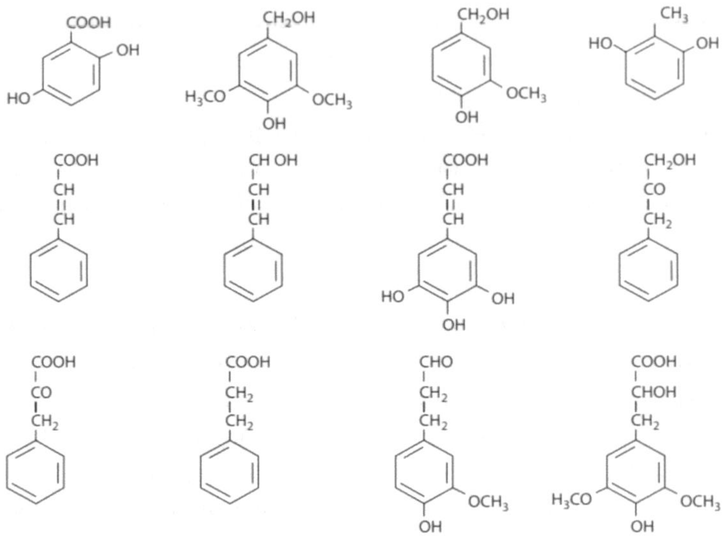

Figure 7.3. *Representation of the products of performic acid oxidizing coal.*

Source: Raj, Creative Commons License.

7.3. BACTERIAL OXIDATION OF COAL

The issue of stream contamination resulting from drainage from acid mines has instigated numerous research endeavors. The researchers concur that the origin of the sulfuric acid along with iron contaminants can be attributed to the pyritic substance linked with coal and coal-containing layers.

The process of oxidizing iron disulfides leads to the generation of ferrous sulfate and sulfuric acid (Silverman et al., 1961). The oxidation process that follows the reaction of ferrous sulfate results in the production of further hydrated oxides of iron and sulfuric acid. The latter refers to the formation of visually unappealing muds that exhibit a color spectrum ranging from yellow to red, which is a distinctive feature of streams that receive acid mine drainage (AMD).

Chemoautotrophic microorganisms that oxidize iron and sulfur have been frequently obtained from acidic aquatic environments, both in the United States and elsewhere. These microorganisms have been demonstrated to enhance the pace of pyrite along with other sulfide mineral oxidation (Temple & Delchamps, 1953). The pyrite oxidation was quantified by the workers through two methods: titration for heightened acidity and assessment of the liberation of easily soluble iron or sulfate S from the previously insoluble minerals.

The utilization of autotrophic organisms that oxidize iron has been documented in the elimination of pyrite from coal, as well as in the secondary retrieval of molybdenum and copper from their respective sulfide minerals (Acharya et al., 2001).

In general, it can be observed that the oxidation rate of pyrite is directly proportional to the extent of its exposure, i.e., the available surface area of pyrite. To achieve the maximum exposure of the embedded pyrite, the coal needs to be in a fragmented state. *F. ferrooxidans* has the ability to expedite the oxidation process of various pyritic substances and marcasite with coarse crystalline structure, however, it is incapable of accelerating the oxidation of coarsely crystalline pyrite. The potential impact of *Thiobacillus thiooxidans* on the oxidizing process of the materials being studied is limited, except for marcasite (Rogoff & Wender, 1957). The organism's involvement in pyrite oxidation seems to be negligible. The capacity of test organisms to expedite oxidation is influenced by lattice imperfections or impurities present in the mineral. Coarsely crystalline pyrite exhibited resistance to oxidation, whereas coarsely crystalline marcasite alongside pyrite concretions and concentrates derived from coal were susceptible to oxidation.

7.4. HYDROGENATION

From a historical perspective, the process of coal hydrogenation is not a recent development. In fact, it has been documented for over a century, dating back to the conversion of coal into liquid products through its reaction with hydrogen iodide in enclosed glass tubes at a temperature of 270°C (520°F) (Wang & Astruc, 2015). Furthermore, the process of coal hydrogenation at elevated temperatures and pressures of hydrogen resulted in the development of a technique for the industrial synthesis of liquid fuels from coal via direct hydrogenation. This method ultimately yielded an output of roughly 1,00,000 barrels per day (equivalent to 15.9 million liters per day) of liquid fuels to support the German military campaign during the period spanning from the early to mid-1940s (Beeck, 1950). The process of coal hydrogenation has been regarded as a potential approach for generating liquid fuels, subject to the prevailing demand and the diverse political factors that can impact the contemporary global landscape.

Extensive research has been conducted on the correlation between coal composition and hydrogenation reactivity. It has been observed that coal containing less than 85% w/w carbon (daf) is not a suitable feedstock for liquefaction. The relationship between coal reactivity and rank has been proposed. Coal is an organic substance that lacks hydrogen and has an atomic hydrogen/carbon ratio of around 0.8 (Wang et al., 2012). This is in contrast to crude oils, heavy oil, and bitumen, which have atomic hydrogen/carbon ratios ranging from 1.4:1.8, as shown in Table 7.1 (Chaloner et al., 2013).

The hydrogenating of coal has garnered significant attention due to its potential to produce liquid substitutes for petroleum feedstocks, which explains why it remains a topic of continued interest (Zhou, 2007). The application of chemical hydrogenation in the context of structural analysis has been observed, however, its potential in the realm of coal utilization has been overlooked, unless it can be linked to economic benefits.

The low ratio of hydrogen to carbon atoms in coal necessitates the introduction of hydrogen to facilitate efficient transformation into liquid products. This concept acknowledges the possibility of liquid formation from coal without the need for external hydrogen, as certain components of the coal organic matrix can serve as a hydrogen donor (Brieger & Nestrick, 1974). However, this process typically results in the production of coke with a deficit of hydrogen ($H/C < 0.5$) in significant amounts.

Table 7.1. Comparison of the Hydrogen to Carbon Ratios in Various Fuels

Fuel	**C**	**H**	**O**	**N**	**S**	**H/Ca (a = Atomic Ratio)**
Coals						
Low-volatile bituminous	90.5	4.7	2.8	1.3	0.7	–
Medium-volatile bituminous	88.4	5.0	4.1	1.7	0.8	0.68
High-volatile bituminous	76.4	5.3	15.8	1.6	0.9	0.83
Subbituminous A	75.8	5.3	15.5	1.9	1.5	0.84
Subbituminous B	75.3	5.1	17.4	1.5	0.7	0.81
Subbituminous C	73.7	5.3	19.1	1.3	0.6	0.86
Lignite	70.2	5.6	20.8	1.4	2.0	0.96
Wood	49.9	6.0	43.9	0.2	–	1.44
Bitumen						
Coldlake	83.7	10.4	1.0	0.2	4.7	1.49

Athabasca	83.1	10.3	1.6	0.4	4.6	1.49
Crude oil	85.8	13.0	–	0.2	1.0	1.82
Gasoline	86.0	14.0	–	–	–	1.95
Natural gas	75.8	24.2	–	–	–	3.83

Source: John F. Fryer, Creative Commons License.

It is noteworthy to mention that the process of water formation through the chemical reaction of hydrogen with oxygen species in coal is deemed inefficient with respect to hydrogen consumption by several researchers. Consequently, alternative methods for coal pretreatment that eliminate some or all of the oxygen before hydrogenation to liquids are deserving of contemplation. However, there could be potential advantages associated with such a therapeutic approach.

There exist certain observations indicating that noncaking coals can develop coking properties upon being subjected to hydrogen heating at temperatures slightly below the point of active decomposition, typically around 350°C (660°F) (Halpern, 1982). The acquired coking properties were attributed to the depletion of oxygen and various other functional groups, along with the inclusion of hydrogen into the coal. Chemical dehydrogenation seems to exhibit a reversal of this phenomenon and also serves as a complete inhibitor of tar production.

The process of fully oxidizing coal can be symbolized by its transformation into carbon dioxide (Elias et al., 2009).

$$C + O_2 = CO_2$$

The complete hydrogenation of coal can (in theory) be represented by conversion to methane:

$$C + 2H_2 ® CH_4$$

It is widely recognized that the process of coal hydrogenation does not involve the complete conversion of coal to methane through gasification, as per the present circumstances. The process involves various techniques that impact the increase of the intrinsic atomic ratio of hydrogen and carbon, leading to the production of partially hydrogenated byproducts such as tar or even more fluid volatile products.

The process of transforming coal into liquid hydrocarbons necessitates a significant quantity of hydrogen, that can only be effectively introduced at temperatures surpassing 350°C (660°F) (Uematsu et al., 1996). Typically, when coal is subjected to decomposition at temperatures less than 350°C

(<660°F), the process of hydrogenation does not proceed further than the point where the coal is transformed into a product that exhibits a notable increase in pyridine solubility.

Given's (1984) research highlights a feasible method of conversion that does not necessitate substantial coal decomposition. This approach involves the reaction of coal with lithium in a solvent, such as ethylenediamine, at a temperature of around 100°C (212°F). Hence, it is feasible to incorporate a maximum of 55 additional hydrogen atoms per 100 carbon atoms into coal utilizing this technique, despite the intricacies linked with the integration of solvent into the end product. The properties and ultimate fate of the product are significantly influenced by the rank of coal and the conditions of the reaction, similar to the oxidation of coal. Hydrogenation yields a diverse range of products, including heavy carbonaceous substances such as cokes, chars, tars, and heavy oils, as well as lighter fractions like naphtha and even methane (Blaser et al., 2003). Under moderate reaction conditions, asphalts tend to be the primary output, whereas under more severe reaction conditions, liquids with boiling points similar to those of petroleum naphtha (30°C–200°C [85°F–390°F]) are generated. In instances of highly rigorous circumstances, methane emerges as the principal output, leading to the process being commonly denoted as hydrogasification.

The process of hydrogenation of coal can be broadly classified into two categories, namely, liquid-phase hydrogenation and vapor-phase hydrogenation, with a focus on coal utilization.

7.4.1. Liquid-Phase Hydrogenation

The process of hydrogenating coal in a liquid phase has been a subject of significant interest since the early stages of research, ultimately leading to one of the earliest commercial advancements in the production of liquid fuels derived from coal (Iwasaki et al., 2014). Recently, there has been a shift in research focus towards exploring cost-effective alternatives for hydrogen production, developing more efficient catalysts, optimizing operating conditions to reduce severity, and streamlining operational protocols. There has been a notable focus on removing the carrier liquid and achieving a simpler separation of the resulting products.

The process of hydrogenating coal in the liquid phase is feasible within the temperature range of 370°C to 540°C (700°F to 1,000°F) and under pressure conditions ranging from 1,500 to 7,500 psi (Jessop et al., 1995). Elevating the process temperature typically results in a rise in the production of naphtha and gases, while potentially decreasing the production rate of middle distillates. However, it is uncommon for this temperature increase to reduce the yield of heavier products. The duration of residence (or reaction) may be approximately 2 hours, and prolonging this duration would not only enhance the production of gas, naphtha, and middle distillate, but also result in a reduction in the yield of heavier oils. The overall conversion can be significantly impacted by an elevation in hydrogen pressure, for instance, from 1,200 to 1,800 psi, and may potentially result in a twofold increase in the conversion yield, given certain optimal conditions. Nevertheless, additional factors, such as the distribution of products, may render the heightened conversion disadvantageous and, consequently, not a genuine advantage to the process.

Considerable attention has been devoted to the chemistry underlying the liquid-phase hydrogenating of coal (Wollenburg et al., 2019). It has been posited that the early stages of this process involve the formation of hydroaromatic compounds through the interaction of hydrogen with the constituents of the liquid medium. The hydroaromatic compounds provide hydrogen to the hydrogen-deficient groups that arise during coal decomposition. This reaction, like several other coal reactions, can only occur efficiently at temperatures exceeding 350°C (660°F). It has been hypothesized that intricate free-radical progressions play a role in the stabilization of coal fragments through the process of hydrogen abstraction from the donor vehicle, without the need for the catalysts or elevated pressures. The utilization of tetrahydronaphthalene (tetralin) as a hydrogen donor in kinetic studies has revealed that coal liquefaction can be achieved with adequate hydrogen supply when the ratio of vehicle coal is approximately 2:1 (Shimizu et al., 2007). Moreover, the conversion process occurs expeditiously at temperatures around 400°C (750°F). Under the specified conditions illustrated in Figures 7.5 and 7.6, it can be observed that coal achieves near-complete solubility in pyridine within a duration of 10 minutes, without any significant rise in the level of hydrogen of the resultant product as compared to the initial coal (Knowles, 1983) (Table 7.2 and Figure 7.4).

Table 7.2. Representation of Catalysts for Liquid-Phase Hydrogenation of Coal

Catalyst	Comments
Tin oxalate + chlorides	Superior to iron; expensive and corrosive.

Ammonium molybdate	Excellent catalyst yielding low-ash oil; expensive without recovery.
Zinc chloride	Produces more aliphatic product; high proportions required Co–Mo.
(“moly”) on silica alumina	Promotes hydrodesulfurization with liquefaction.

Source: James G. Speight, Creative Commons License.

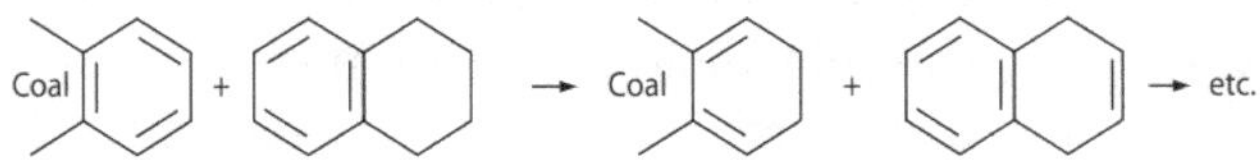

Figure 7.4. *Representation of hydrogen transfer reactions.*

Source: Marcel Dekker, Creative Commons License.

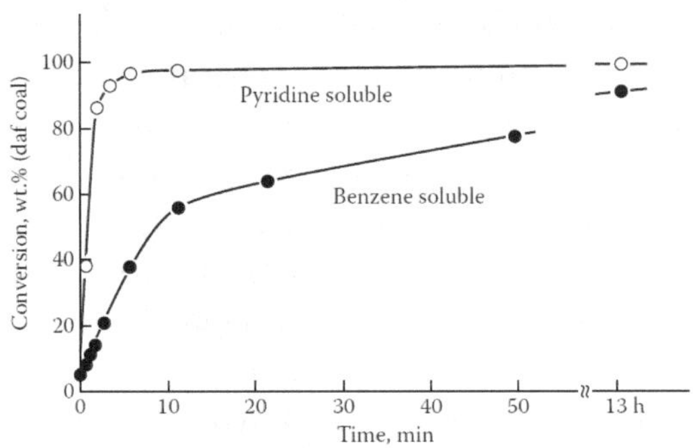

Figure 7.5. *Variation of conversion of high volatile C bituminous coal with time.*

Source: Neavel, Creative Commons License.

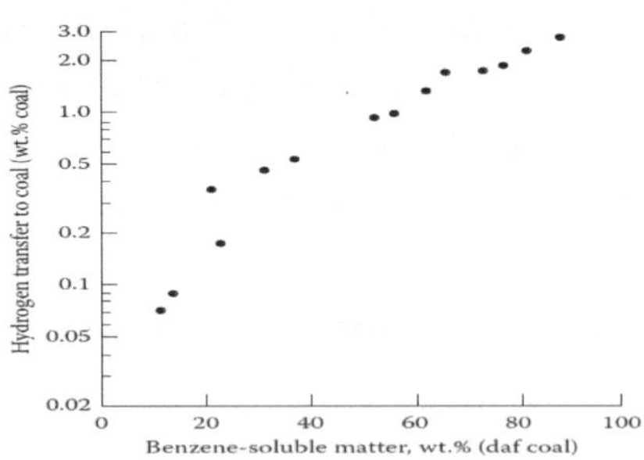

Figure 7.6. *Variation of conversion of hvC bituminous coal with hydrogen transfer.*

Source: Neavel, Creative Commons License.

Coal samples that fall within the high-volatile C bituminous coal category and below exhibit comparable reaction rates and tend to disintegrate within the donor vehicle at the temperature of reaction (Knowles & Noyori, 2007). This characteristic reduces the need for pre-crushing of the coal prior to reaction. It is important to bear in mind that coal is comprised of various macerals, and the presence of different macerals may have an unforeseen impact on the efficacy of the process. In instances where vitrinites and exinites are present, they may undergo complete reaction. However, fusinites tend to be inert and can be observed in the residue alongside the mineral matter.

The maceral content of coal is a crucial factor to consider in addition to the widely accepted principle that coal rank has a significant impact on coal conversion.

7.4.2. Vapor-Phase Hydrogenation

The process of vapor-phase hydrogenation of coal involves the direct reaction between coal and hydrogen, without the use of any intermediary substance (Wang et al., 2011). The reaction may necessitate elevated temperatures or may occur under reduced pressures compared to those utilized in the coal liquid-phase hydrogenation process.

Evidently, the reaction's characteristics in the absence of a carrier will deviate from those observed in an existence of one, due to the direct interaction between hydrogen and coal. Furthermore, it is noteworthy that the process of secondary hydrogenation, also known as vapor-phase hydrogenation, may ensue subsequent to the primary hydrogenation. This involves the volatile byproducts resulting from coal decomposition or the reaction of coal with hydrogen undergoing further reaction with hydrogen to alter the composition of the primary reaction products (Jessop et al., 2004).

A preliminary hydrogenation process may be utilized as an additional stage, which can effectively reduce the levels of unsaturated compounds and oxygen species that have the potential to significantly diminish the lifespan of the catalyst (Ito et al., 2014).

Coals of lower rank exhibit a greater propensity for reactivity in comparison to coals of higher rank or coke. The production of methane can occur through two pathways: the direct reaction of coal carbon and hydrogen, and the thermal decomposition and hydrogenation (hydrocracking) of higher-molecular-weight compounds that are formed during the initial stages of hydrogenation (Glorius, 2005). The general methane production rate can be considered as first-order in relation to the partial pressure of hydrogen.

Hydrogasification reactions exhibit a preference for methane conversion under conditions of elevated pressure and relatively lower temperatures. The reaction exhibits a swift pace when exposed to temperatures exceeding 480°C (900°F) (Shultz & Krska, 2007). Although the initial coal is typically more responsive than the resultant char, it may seem counterintuitive that the char can frequently undergo conversion to methane with greater ease than coal. There are also reports that sodium carbonate can increase the rate of the reaction of coal with hydrogen in the range 815°C–900°C (1,500°F–1,650°F) (Urban et al., 2012). Nonetheless, tin(II) chloride has been observed to have remarkably advantageous impacts for the catalysts that are introduced to the reaction mixture in the form of a powder.

In contrast, it has been observed that ammonium molybdate, tin(II) chloride, nickel(II) chloride, and iron(II) sulfate are advantageous catalysts when incorporated into the reaction mixture through solution impregnation of the coal.

7.5. HALOGENATION

The reactivity of coal towards halogens has been observed and documented in scientific literature. The halogenation of coal was initially demonstrated through the treatment of cannel coal with "chlorine water," resulting in the production of alkali-soluble products (Neumann et al., 2008). Additionally, it was observed that bituminous coal could also be made soluble through cure with hydrochloric acid and potassium chlorate.

Further research has indicated that chlorination can be accomplished through various methods, including the exposure of pulverized coal to chlorine, the passage of chlorine gases by a solution of coal in water or carbon tetrachloride, and the reaction of coal with chlorine dioxide in a vessel with pressure. The solubility of the products resulting from the chlorination of lower-rank coals may vary depending on the specific procedure employed, with some exhibiting alkali solubility (Schnelle & Strimple, 1995).

The chlorine content in the resulting products from each of these methods exhibits a range of values, with lignite yielding the highest amount at approximately 40% w/w and bituminous coal producing a lower amount at around 25% w/w (Tang et al., 2018). It is possible that the reaction temperature could experience an elevation. The exothermic process of halogenation of desiccated coal using gaseous chlorine results in the liberation of hydrogen chloride. Upon combustion of the chlorinated substance, there is a negligible amount of tar produced. The aforementioned observations have been postulated to be explicable through the mechanism of concurrent addition and substitution reactions (as illustrated in Figure 7.7) whereby chlorine is incorporated into the coal.

The chemistry of chlorination in relation to coal is a complex matter, with evidence suggesting that the characteristics of chlorinated products are influenced not only by the degree of chlorination, but also by the type of coal and the reagent used (Podgoršek et al., 2009). As previously stated, it has been suggested that the dissimilarities among coals are manifested in an elevated absorption of chlorine in lignite compared to coals of higher rank.

The results of attempts to chlorinate various macerals suggest that exinite exhibits a higher degree of disordered structure compared to fusinite, and that the quantity of condensed aromatic rings that are present does not exceed two (Ball & Mitchell, 1978). Furthermore, sulfur exhibits stronger binding affinity and possesses a higher count of peripheral groups. Conversely, fusinite exhibits greater levels of structural density, with a minimal quantity of hydroaromatic structures and a prevalence of condensed rings numbering at least two.

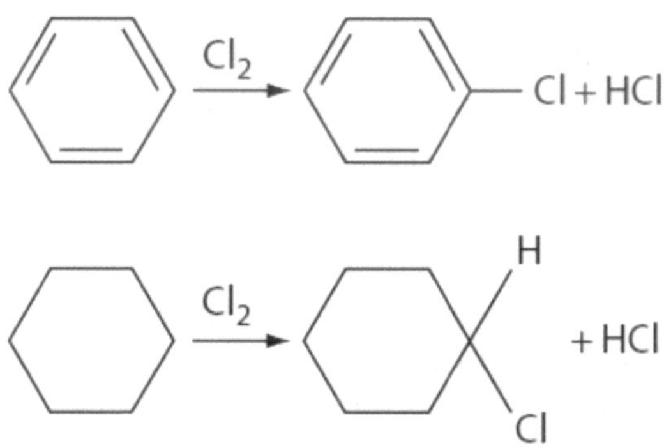

Figure 7.7. *Illustration of chlorination (substitution) reactions.*

Source: Taylor, Creative Commons License.

The chemical reactions between coal and bromine, iodine, and fluorine have all been studied to a lesser extent compared to the reaction with chlorine. The utilization of sulfur tetrafluoride for the purpose of introducing fluorine into particular sites of coal has been documented as a dependable method in the literature. This process involves the reaction of lignite with sulfur tetrafluoride. Bromination, a chemical process that involves trembling powdered coal with bromine water in carbon tetrachloride or chloroform, is thought to entail both addition and substitution reactions. The resulting products may contain as much as 20% w/w bromine (Nebergall, 2014). Analogous to the aforementioned notion, the interplay between iodine and coal has been postulated. However, the reaction between coal and fluorine will culminate in a conglomeration of fluorocarbon oils.

Coal can undergo a reaction with mixed halogens, such as chlorine trifluoride, under specific temperature conditions ranging from 205°C to 400°C (400°F to 750°F). This reaction results in the production of chlorofluorocarbon oils alongside gases (Meshram et al., 2005). It is possible that the oils, which are the previous products, may contain a significant portion (approximately 60%) of the halogen that has undergone a reaction.

7.6. ALKYLATION

The aromatic qualities of coal, which is a multidimensional macromolecule, are the primary factor that distinguish it from other forms of carbon. Alkylation processes, in which a group of alkyl atoms substitutes an atom of hydrogen in an aromatic ring, may be carried out on it successfully. Coal may have its reactivity toward liquefaction increased by undergoing a process known as an alkylation (Ettinger et al., 1986). This process has been the subject of substantial research as a potential method for increasing coal's volatility in a variety of organic solvents.

Coal's solubility may be improved by exposing it to a process with an alkyl chloride in the presence of aluminum chloride and making use of a solvent such as carbon disulfide. This reaction should take place in the presence of aluminum chloride. This technique has the potential to result in a substantial rise in the proportion of pyridine-soluble material found in the coal, from roughly 11% in the initial coal to as high as 39% (Larsen & Kuemmerle, 1976).

The Friedel-Crafts reaction of coal is a process that involves several stages that is not just an alkylation in its purest form. The circumstances under which this reaction occurs are called Friedel-Crafts conditions (Liotta et al., 1981). The chemical composition of coal molecules and the mineral content of coal both have the ability to either inhibit or promote the processes described above. The usage of extraction allows for the evaluation of the effectiveness of reactions such as the degradation of coal, the disintegration of coal, and the transalkylation of coal. Under Friedel-Crafts conditions, the interaction between coal and alkylating chemicals is a complicated chain reaction that comprises alkylation, transalkylation, dehydrogenation, degradation, and dissociation. All of these events take place simultaneously.

Additional methods for the alkylation process include the use of alcohols as alkylating agents in combination with the use of hydrogen fluoride and the zinc chloride-alcohol system, as well as the inclusion of an alkyl iodide when cadmium is present in the environment (Liotta & Brons, 1981). Both of these strategies may be found in the zinc chloride-alcohol system.

Because the product of the reaction may display a substantially higher degree of aromaticity in comparison to the original coal, the reactions that are involved in the process are significantly more complicated when compared to the conventional alkylation. As shown in Figure 7.8, the emergence of such a phenomena may result from the contemporaneous occurrence of a condensation process that is aided by the existence of aluminum chloride (Liotta, 1979). This process is represented in the figure. Another possible origin is the condensation and following aromatization of native alkyl side chains. This would be an alternative origin. It is important to take into consideration the fact that aluminum chloride has been shown to hasten the synthesis of graphite from aromatic compounds. This information is essential to the discussion at hand.

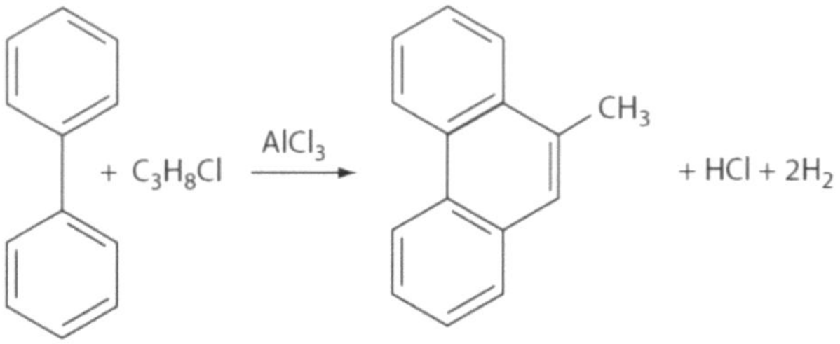

Figure 7.8. *The Friedel–Crafts cycloalkylation in conjunction with aromatization.*

Source: John F. Fryer, Creative Commons License.

The phenomenon of coal alkylation has been extended to encompass a diverse range of investigations pertaining to the structural characteristics of coal. The process described can be classified as a form of "reductive alkylation" (Sternberg et al., 1971). The initial stage of this process involves the reaction of coal alongside metallic potassium and naphthalene in tetrahydrofuran (THF). By means of an electron transfer mechanism, a coal anion is generated, which subsequently undergoes alkylation via an alkyl halide. The initial conception of the reductive alkylation of coal postulated a process involving uncomplicated chemical reactions with negligible occurrence of side reactions (Sternberg & Delle Donne, 1974). However, numerous subsequent researchers have raised significant inquiries regarding the credibility of the originally proposed reaction mechanism. The occurrence of intricate side reactions raises uncertainties regarding both the exact composition of the resulting products and the fundamental structure of the reaction.

7.7. SUMMARY

Chemical reactivity refers to the ability of coal to undergo various chemical reactions, both under normal conditions and at elevated temperatures. The most well-known chemical reaction involving coal is combustion, which occurs when coal reacts with oxygen in the presence of heat. Coal combustion and processing can result in the emission of pollutants such as sulfur dioxide and nitrogen oxides. It is worth noting that chemical reactivity of coal can vary depending on its rank, which is determined by factors such as its carbon content, heating value and degree of metamorphism. Different ranks of coal exhibit different reactivity and properties.

REFERENCES

1. Acharya, C., Kar, R. N., & Sukla, L. B., (2001). Bacterial removal of sulfur from three different coals. *Fuel, 80*(15), 2207–2216.
2. Arenillas, A., Pevida, C., Rubiera, F., & Pis, J. J., (2003). Comparison between the reactivity of coal and synthetic coal models☆. *Fuel, 82*(15–17), 2001–2006.
3. Arisoy, A., & Beamish, B., (2015). Mutual effects of pyrite and moisture on coal self-heating rates and reaction rate data for pyrite oxidation. *Fuel, 139*(1), 107–114.
4. Astbury, G. R., & Hawksworth, S. J., (2007). Spontaneous ignition of hydrogen leaks: A review of postulated mechanisms. *International Journal of Hydrogen Energy, 32*(13), 2178–2185.
5. Ball, D. R., & Mitchell, S. C., (1978). Preparation and characterization of a chlorinated low-rank bituminous coal. *Fuel, 57*(6), 365–371.
6. Barrera-Díaz, C., Bilyeu, B., Roa, G., & Bernal-Martinez, L., (2011). Physicochemical aspects of electrocoagulation. *Separation and Purification Reviews, 40*(1), 1–24.
7. Beccalli, E. M., Broggini, G., Martinelli, M., & Sottocornola, S., (2007). C–C, C–O, C–N bond formation on sp^2 carbon by Pd(II)-catalyzed reactions involving oxidant agents. *Chemical Reviews, 107*(11), 5318–5365.
8. Beeck, O., (1950). Hydrogenation catalysts. *Discussions of the Faraday Society, 8*(1), 118–128.
9. Blaser, H. U., Malan, C., Pugin, B., Spindler, F., Steiner, H., & Studer, M., (2003). Selective hydrogenation for fine chemicals: Recent trends and new developments. *Advanced Synthesis and Catalysis, 345*(1–2), 103–151.
10. Blomqvist, P., & Persson, B., (2003). *Spontaneous Ignition of Biofuels – A Literature Survey of Theoretical and Experimental Methods* (Vol. 1, pp. 4–7).
11. Brieger, G., & Nestrick, T. J., (1974). Catalytic transfer hydrogenation. *Chemical Reviews, 74*(5), 567–580.
12. Burger, J. G., (1976). Spontaneous ignition in oil reservoirs. *Society of Petroleum Engineers Journal, 16*(02), 73–81.
13. Chaloner, P. A., Esteruelas, M. A., Joó, F., & Oro, L. A., (2013). *Homogeneous Hydrogenation* (Vol. 15, pp. 5–10). Springer Science & Business Media.
14. Chan, M. L., Jones, J. M., Pourkashanian, M., & Williams, A., (1999).

The oxidative reactivity of coal chars in relation to their structure. *Fuel, 78*(13), 1539–1552.

15. Chaston, J. C., (1964). Reaction of oxygen with the platinum metals: Part I: The oxidation of platinum. *Platinum Metals Review, 8*(2), 50–54.
16. Clark, C. S., (1966). Oxidation of coal mine pyrite. *Journal of the Sanitary Engineering Division, 92*(2), 127–146.
17. Clerici, M. G., & Ingallina, P., (1998). Oxidation reactions with *in situ* generated oxidants. *Catalysis Today, 41*(4), 351–364.
18. Crelling, J. C., Skorupska, N. M., & Marsh, H., (1988). Reactivity of coal macerals and lithotypes. *Fuel, 67*(6), 781–785.
19. Cybulska, M., Jedraszek-Bomba, A., Kuberski, S., & Wrzosek, H., (2008). Methods of chemical and physicochemical analysis in the identification of archaeological and historical textiles. *Fibers and Textiles in Eastern Europe, 5*(70), 67–73.
20. Deng, J., Ma, X., Zhang, Y., Li, Y., & Zhu, W., (2015). Effects of pyrite on the spontaneous combustion of coal. *International Journal of Coal Science and Technology, 2*, 306–311.
21. dos Santos, R. P., Martins, J., Gadelha, C., Cavada, B., Albertini, A. V., Arruda, F., & Freire, V., (2014). Coal fly ash ceramics: Preparation, characterization, and use in the hydrolysis of sucrose. *The Scientific World Journal, 1*, pp. 3–8.
22. Dryer, F. L., Chaos, M., Zhao, Z., Stein, J. N., Alpert, J. Y., & Homer, C. J., (2007). Spontaneous ignition of pressurized releases of hydrogen and natural gas into air. *Combustion Science and Technology, 179*(4), 663–694.
23. Elias, D. C., Nair, R. R., Mohiuddin, T. M. G., Morozov, S. V., Blake, P., Halsall, M. P., & Novoselov, K. S., (2009). Control of graphene's properties by reversible hydrogenation: Evidence for graphane. *Science, 323*(5914), 610–613.
24. Ettinger, M., Nardin, R., Mahasay, S. R., & Stock, L. M., (1986). An investigation of the O-and C-alkylation of coal. *The Journal of Organic Chemistry, 51*(14), 2840–2842.
25. Freeman, G., & Lefebvre, A. H., (1984). Spontaneous ignition characteristics of gaseous hydrocarbon-air mixtures. *Combustion and Flame, 58*(2), 153–162.
26. Fung, D. P., & Kim, S. D., (1983). Laboratory gasification study of Canadian coals: Part 2: Chemical reactivity and coal rank. *Fuel, 62*(11), 1337–1340.
27. Garrels, R. M., Mackenzie, F. T., & Hunt, C., (1973). *Chemical Cycles and*

the Global Environment: Assessing Human Influences (Vol. 1, pp. 6–9).

28. Glorius, F., (2005). Asymmetric hydrogenation of aromatic compounds. *Organic and Biomolecular Chemistry, 3*(23), 4171–4175.
29. Good, A., Durden, D. A., & Kebarle, P., (1970). Mechanism and rate constants of ion-molecule reactions leading to formation of $H^+ (H_2O)$ n in moist oxygen and air. *The Journal of Chemical Physics, 52*(1), 222–229.
30. Gray, B. F., Griffiths, J. F., & Hasko, S. M., (1984). Spontaneous ignition hazards in stockpiles of cellulosic materials: Criteria for safe storage. Journal of chemical technology and biotechnology. *Chemical Technology, 34*(8), 453–463.
31. Griffiths, J. F., & Scott, S. K., (1987). Thermokinetic interactions: Fundamentals of spontaneous ignition and cool flames. *Progress in Energy and Combustion Science, 13*(3), 161–197.
32. Guy, R. H., & Hadgraft, J., (1988). Physicochemical aspects of percutaneous penetration and its enhancement. *Pharmaceutical Research, 5*(1), 753–758.
33. Halliwell, B., Hoult, J. R., & Blake, D. R., (1988). Oxidants, inflammation, and anti-inflammatory drugs. *The FASEB Journal, 2*(13), 2867–2873.
34. Halpern, J., (1982). Mechanism and stereoselectivity of asymmetric hydrogenation. *Science, 217*(4558), 401–407.
35. Hummelshøj, J. S., Blomqvist, J., Datta, S., Vegge, T., Rossmeisl, J., Thygesen, K. S., & Nørskov, J. K., (2010). Communications: Elementary oxygen electrode reactions in the aprotic Li-air battery. *The Journal of Chemical Physics, 132*(7), 071101.
36. Ito, J. I., & Nishiyama, H., (2014). Recent topics of transfer hydrogenation. *Tetrahedron Letters, 55*(20), 3133–3146.
37. Iwasaki, K., Wan, K. K., Oppedisano, A., Crossley, S. W., & Shenvi, R. A., (2014). Simple, chemoselective hydrogenation with thermodynamic stereocontrol. *Journal of the American Chemical Society, 136*(4), 1300–1303.
38. Jackson, J. L., (1951). Spontaneous ignition temperatures. *Industrial and Engineering Chemistry, 43*(12), 2869–2870.
39. Jessop, P. G., Ikariya, T., & Noyori, R., (1995). Homogeneous hydrogenation of carbon dioxide. *Chemical Reviews, 95*(2), 259–272.
40. Jessop, P. G., Joó, F., & Tai, C. C., (2004). Recent advances in the homogeneous hydrogenation of carbon dioxide. *Coordination Chemistry Reviews, 248*(21–24), 2425–2442.
41. Jiménez-Escrig, A., & Sánchez-Muniz, F. J., (2000). Dietary fiber from

edible seaweeds: Chemical structure, physicochemical properties and effects on cholesterol metabolism. *Nutrition Research, 20*(4), 585–598.
42. Jörissen, L., (2006). Bifunctional oxygen/air electrodes. *Journal of Power Sources, 155*(1), 23–32.
43. Kaufman, F., (1958). The air afterglow and its use in the study of some reactions of atomic oxygen. *Proceedings of the Royal Society of London: Series A; Mathematical and Physical Sciences, 247*(1248), 123–139.
44. Keenan, C. R., & Sedlak, D. L., (2008). Factors affecting the yield of oxidants from the reaction of nanoparticulate zero-valent iron and oxygen. *Environmental Science and Technology, 42*(4), 1262–1267.
45. Knowles, W. S., & Noyori, R., (2007). Pioneering perspectives on asymmetric hydrogenation. *Accounts of Chemical Research, 40*(12), 1238–1239.
46. Knowles, W. S., (1983). Asymmetric hydrogenation. *Accounts of Chemical Research, 16*(3), 106–112.
47. Kogelschatz, U., Eliasson, B., & Hirth, M., (1988). *Ozone Generation from Oxygen and Air: Discharge Physics and Reaction Mechanisms* (Vol. 1, pp. 5–9).
48. Kurose, R., Tsuji, H., & Makino, H., (2001). Effects of moisture in coal on pulverized coal combustion characteristics. *Fuel, 80*(10), 1457–1465.
49. Kuznetsov, Y. I., & Kazansky, L. P., (2008). Physicochemical aspects of metal protection by azoles as corrosion inhibitors. *Russian Chemical Reviews, 77*(3), pp. 215–219.
50. Larsen, J. W., & Kuemmerle, E. W., (1976). Alkylation and depolymerization reactions of coal: A selective review with supplementary experiments. *Fuel, 55*(3), 162–169.
51. Li, Y., Yao, Y., Liu, X., Sun, H., & Ni, W., (2013). Improvement on pozzolanic reactivity of coal gangue by integrated thermal and chemical activation. *Fuel, 109*, 527–533.
52. Li, Z., Xu, G., & Shi, X., (2021). Reactivity of coal fly ash used in cementitious binder systems: A state-of-the-art overview. *Fuel, 301*(1), 121031.
53. Lieberman, H., & Vemuri, N. M., (2015). Chemical and physicochemical approaches to solve formulation problems. In: *The Practice of Medicinal Chemistry* (Vol. 1, pp. 767–791). Academic Press.
54. Liotta, R., & Brons, G., (1981). Coal. Kinetics of O-alkylation. *Journal of the American Chemical Society, 103*(7), 1735–1742.
55. Liotta, R., (1979). Selective alkylation of acidic hydroxyl groups in coal.

Fuel, 58(10), 724–728.

56. Liotta, R., Rose, K., & Hippo, E., (1981). O-Alkylation chemistry of coal and its implications for the chemical and physical structure of coal. *The Journal of Organic Chemistry, 46*(2), 277–283.
57. Littman, F. E., Church, F. M., & Kinderman, E. M., (1961). A study of metal ignitions I. The spontaneous ignition of titanium. *Journal of the Less Common Metals, 3*(5), 367–378.
58. Mareschal, L., Bonnaud, P., Turpault, M. P., & Ranger, J., (2010). Impact of common European tree species on the chemical and physicochemical properties of fine earth: An unusual pattern. *European Journal of Soil Science, 61*(1), 14–23.
59. Marnett, L. J., Riggins, J. N., & West, J. D., (2003). Endogenous generation of reactive oxidants and electrophiles and their reactions with DNA and protein. *The Journal of Clinical Investigation, 111*(5), 583–593.
60. Meshram, H. M., Reddy, P. N., Sadashiv, K., & Yadav, J. S., (2005). Amberlyst-15®-promoted efficient 2-halogenation of 1, 3-keto-esters and cyclic ketones using N-halosuccinimides. *Tetrahedron Letters, 46*(4), 623–626.
61. Mishra, B., Hassan, P. A., Priyadarsini, K. I., & Mohan, H., (2005). Reactions of biological oxidants with selenourea: formation of redox active nanoselenium. *The Journal of Physical Chemistry B, 109*(26), 12718–12723.
62. Molina, A., & Mondragon, F., (1998). Reactivity of coal gasification with steam and CO_2. *Fuel, 77*(15), 1831–1839.
63. Mondragon, F., Itoh, H., & Ouchi, K., (1982). Solubility increase of coal by alkylation with various alcohols. *Fuel, 61*(11), 1131–1134.
64. Nassimbeni, L. R., (2003). Physicochemical aspects of host–guest compounds. *Accounts of Chemical Research, 36*(8), 631–637.
65. Nebergall, R., (2014). Halogenated carbon sorbents. *Mercury Control: For Coal-Derived Gas Streams*, *1*, pp. 311–322.
66. Neumann, C. S., Fujimori, D. G., & Walsh, C. T., (2008). Halogenation strategies in natural product biosynthesis. *Chemistry and Biology, 15*(2), 99–109.
67. Osório, E., Gomes, M. D. L. I., Vilela, A. C., Kalkreuth, W., de Almeida, M. A. A., Borrego, A. G., & Alvarez, D., (2006). Evaluation of petrology and reactivity of coal blends for use in pulverized coal injection (PCI). *International Journal of Coal Geology, 68*(1–2), 14–29.
68. Pietrzak, R., Grzybek, T., & Wachowska, H., (2007). XPS study of pyrite-

free coals subjected to different oxidizing agents. *Fuel, 86*(16), 2616–2624.

69. Ping, N., Lili, Y. U., Honghong, Y. I., Xiaolong, T. A. N. G., Hua, L. I., Hongyan, W., & Lina, Y., (2010). Effect of Fe/Cu/Ce loading on the coal-based activated carbons for hydrolysis of carbonyl sulfide. *Journal of Rare Earths, 28*(2), 205–210.
70. Podgoršek, A., Zupan, M., & Iskra, J., (2009). Oxidative halogenation with "green" oxidants: Oxygen and hydrogen peroxide. *Angewandte Chemie International Edition, 48*(45), 8424–8450.
71. Ritchie, A. G., (1981). A review of the rates of reaction of uranium with oxygen and water vapor at temperatures up to 300°C. *Journal of Nuclear Materials, 102*(1–2), 170–182.
72. Rogoff, M. H., & Wender, I., (1957). The microbiology of coal I: Bacterial oxidation of phenanthrene. *Journal of Bacteriology, 73*(2), 264–268.
73. Rokosov, Y. V., Bodoev, N. V., & Sidel'nikov, V. N., (1992). Organic constituents of sapropelitic coal: recovery by high-temperature alkaline hydrolysis and characterization. *Fuel, 71*(10), 1161–1168.
74. Salinger, A. G., Aris, R., & Derby, J. J., (1994). Modeling the spontaneous ignition of coal stockpiles. *AIChE Journal, 40*(6), 991–1004.
75. Santelli, C. M., Pfister, D. H., Lazarus, D., Sun, L., Burgos, W. D., & Hansel, C. M., (2010). Promotion of Mn(II) oxidation and remediation of coal mine drainage in passive treatment systems by diverse fungal and bacterial communities. *Applied and Environmental Microbiology, 76*(14), 4871–4875.
76. Schnelle, R. C., & Strimple, P. D., (1995). *A Preliminary Study on the Effects of Combined Halogenation and Thermal Recirculation at a Coal-Fired Generating Station (No. EPRI-TR-105116; CONF-9502124-)* (Vol. 3, No. 1, pp. 3–9). Electric Power Research Inst. (EPRI), Palo Alto, CA (United States).
77. Schwitzgebel, J., Ekerdt, J. G., Gerischer, H., & Heller, A., (1995). Role of the oxygen molecule and of the photogenerated electron in TiO_2-photocatalyzed air oxidation reactions. *The Journal of Physical Chemistry, 99*(15), 5633–5638.
78. Shao, S., Ma, B., Wang, C., & Chen, Y., (2023). Thermal behavior and chemical reactivity of coal gangue during pyrolysis and combustion. *Fuel, 331*(1), 125–927.
79. Shultz, C. S., & Krska, S. W., (2007). Unlocking the potential of asymmetric hydrogenation at Merck. *Accounts of Chemical Research, 40*(12), 1320–1326.

80. Silverman, M. P., Rogoff, M. H., & Wender, I., (1961). Bacterial oxidation of pyritic materials in coal. *Applied Microbiology, 9*(6), 491–496.
81. Sternberg, H. W., & Delle Donne, C. L., (1974). Solubilization of coals by reductive alkylation. *Fuel, 53*(3), 172–175.
82. Sternberg, H. W., Delle Donne, C. L., Pantages, P., Moroni, E. C., & Markby, R. E., (1971). Solubilization of an lvb coal by reductive alkylation. *Fuel, 50*(4), 432–442.
83. Suuberg, E. M., Otake, Y., Yun, Y., & Deevi, S. C., (1993). Role of moisture in coal structure and the effects of drying upon the accessibility of coal structure. *Energy and Fuels, 7*(3), 384–392.
84. Swern, D., Findley, T. W., Billen, G. N., & Scanlan, J. T., (1947). Determination of oxirane oxygen. *Analytical Chemistry, 19*(6), 414–415.
85. Tadema, H. J., & Wiejdema, J., (1970). Spontaneous ignition of oil sands. *Oil Gas J. (United States)*, *68*(50), 5–8.
86. Tang, R. J., Milcent, T., & Crousse, B., (2018). Regioselective halogenation of arenes and heterocycles in hexafluoroisopropanol. *The Journal of Organic Chemistry, 83*(2), 930–938.
87. Temple, K. L., & Delchamps, E. W., (1953). Autotropic bacteria and the formation of acid in bituminous coal mines. *Applied Microbiology, 1*(5), 255–258.
88. Traylor, T. G., Kim, C., Richards, J. L., Xu, F., & Perrin, C. L., (1995). Reactions of iron(III) porphyrins with oxidants. Structure-reactivity studies. *Journal of the American Chemical Society, 117*(12), 3468–3474.
89. Uematsu, N., Fujii, A., Hashiguchi, S., Ikariya, T., & Noyori, R., (1996). Asymmetric transfer hydrogenation of imines. *Journal of the American Chemical Society, 118*(20), 4916–4917.
90. Urban, S., Beiring, B., Ortega, N., Paul, D., & Glorius, F., (2012). Asymmetric hydrogenation of thiophenes and benzothiophenes. *Journal of the American Chemical Society, 134*(37), 15241–15244.
91. Wachowska, H., Andrzejak, A., & Kozłowski, M., (1988). Reductive alkylation of coals in liquid ammonia. *Fuel Processing Technology, 19*(3), 215–227.
92. Wang, D. S., Chen, Q. A., Lu, S. M., & Zhou, Y. G., (2012). Asymmetric hydrogenation of heteroarenes and arenes. *Chemical Reviews, 112*(4), 2557–2590.
93. Wang, D., & Astruc, D., (2015). The golden age of transfer hydrogenation. *Chemical Reviews, 115*(13), 6621–6686.
94. Wang, W., Wang, S., Ma, X., & Gong, J., (2011). Recent advances in

catalytic hydrogenation of carbon dioxide. *Chemical Society Reviews, 40*(7), 3703–3727.

95. Wen, H., Yu, Z., Deng, J., & Zhai, X., (2017). Spontaneous ignition characteristics of coal in a large-scale furnace: An experimental and numerical investigation. *Applied Thermal Engineering, 114*(1), 583–592.
96. Wen, J. X., Xu, B. P., & Tam, V. H. Y., (2009). Numerical study on spontaneous ignition of pressurized hydrogen release through a length of tube. *Combustion and Flame, 156*(11), 2173–2189.
97. Wollenburg, M., Moock, D., & Glorius, F., (2019). Hydrogenation of borylated arenes. *Angewandte Chemie International Edition, 58*(20), 6549–6553.
98. Xu, B. P., El Hima, L., Wen, J. X., Dembele, S., Tam, V. H. Y., & Donchev, T., (2008). Numerical study on the spontaneous ignition of pressurized hydrogen release through a tube into air. *Journal of Loss Prevention in the Process Industries, 21*(2), 205–213.
99. Yafei, W., Lizhong, Y., Xiaodong, Z., Jiakun, D., Yupeng, Z., & Zhihua, D., (2010). Experiment study of the altitude effects on spontaneous ignition characteristics of wood. *Fuel, 89*(5), 1029–1034.
100. Yeow, J., Chapman, R., Gormley, A. J., & Boyer, C., (2018). Up in the air: oxygen tolerance in controlled/living radical polymerization. *Chemical Society Reviews, 47*(12), 4357–4387.
101. Zabetakis, M. G., Furno, A. L., & Jones, G. W., (1954). Minimum spontaneous ignition temperatures of combustibles in air. *Industrial and Engineering Chemistry, 46*(10), 2173–2178.
102. Zhang, Y., Xu, L., Seetharaman, S., Liu, L., Wang, X., & Zhang, Z., (2015). Effects of chemistry and mineral on structural evolution and chemical reactivity of coal gangue during calcination: Towards efficient utilization. *Materials and Structures, 48*(1), 2779–2793.
103. Zhao, C. X., Liu, J. N., Wang, J., Ren, D., Yu, J., Chen, X., & Zhang, Q., (2021). A ΔE= 0.63 V bifunctional oxygen electrocatalyst enables high-rate and long-cycling zinc–air batteries. *Advanced Materials, 33*(15), 606.
104. Zhou, Y. G., (2007). Asymmetric hydrogenation of heteroaromatic compounds. *Accounts of Chemical Research, 40*(12), 1357–1366.
105. Zou, Y., & Han, B. X., (2001). Preparation of activated carbons from Chinese coal and hydrolysis lignin. *Adsorption Science and Technology, 19*(1), 59–72.

CHAPTER

8 Impacts of Coal Utilization on Health and Safety

Contents

INTRODUCTION

The augmented utilization of coal is linked to various potential impacts on health and safety. The consequences of the range mentioned above, from fatality to premature mortality due to compromised health or exacerbating pre-existing conditions. Photochemical oxidants or radiation can result in genetic defects that may lead to reduced quality of life and deferred losses of life in subsequent generations (Finkelman et al., 2002). Furthermore, non-lethal injuries and illnesses exhibit a spectrum of severity, ranging from acute and critical cases, occasionally accompanied by enduring disability or hospitalization, to postponed illnesses linked solely to stress and primarily manifested as discomforts.

It is important to acknowledge that the discernible negative health impacts easily observable during the 1950s and 1960s due to air pollution are not currently evident at existing air quality levels. The potential negative effects of air pollution on human health, while present, maybe more nuanced, and elusive to identify compared to other health risks. The precise intention of Congress regarding the definition of the term "adverse health effect" remains ambiguous (Mamurekli, 2010). However, the current construal of the legislation suggests that any negative impacts, even those minimal, should be eradicated. The extension of this interpretation may have significant ramifications for forthcoming energy technologies.

The current data about evaluating the harmful impacts of diverse coal technologies on human health is inadequate, inconsistent, and arduous to assess. The reliability of data about certain effects varies significantly. It is a prevalent convention to express the level of certainty linked with documented outcomes by presenting a spectrum of numerical values instead of a solitary figure (Hendryx et al., 2020). The magnitude of ranges varies significantly depending on the type of effect being studied. For instance, the ranges may be relatively narrow, typically a few percent, in cases of accidental deaths resulting from coal mining. Conversely, the ranges may be considerably wider, spanning two orders of magnitude or more, in instances of premature deaths among the general population caused by air pollutant emissions.

The variations in the database can be attributed to several factors, including the clarity of the connection of cause and effect between the impact and its impact on health and safety, the extent of available data on the types and amounts of emissions (especially from novel processes), and the inadequacy of reliable information on human responses to low levels of numerous pollutants (Bezdek, 1993). The primary source for unpredictability in the data utilized for assessing the effects on health pertains to the requisite information for approximating the number of untimely fatalities and incapacities arising from diverse respiratory ailments linked with coal mining (Vamvuka, 1999). An instance of note is the typical latency period of 14 years that transpires between the onset of coming into contact with coal mine dust and the manifestation of clinical indications of coal workers' pneumoconiosis (CWP). Hence, despite the control of current coal dust exposures, the emergence of clinical symptoms of CWP and previous cases of the disease will persist for several years.

Did You Know?

The term "black lung" originated from the appearance of coal miners' lungs affected by coal dust accumulation.

The health and safety implications of coal technologies can be categorized based on their association with different phases of the coal system, their geographical scope (i.e., worldwide, local, or regional), and their impact on occupational workers versus the general public.

8.1. GLOBAL

The primary potential worldwide impact from heightened coal utilization is the prospect of carbon dioxide emissions inducing alterations to the Earth's climate (Balat, 2007). Additional impacts on the well-being of the global populace are expounded upon within this discourse.

In the event of substantial and enduring alterations to the climate due to heightened coal consumption, the potential ramifications for human health and well-being would be contingent upon the degree to which humans can adapt to such changes, as well as the status and prevalence of agricultural practices and living conditions. The possible hazards to human health resulting from alterations in atmospheric carbon dioxide levels and their magnitude remain uncertain (Morris et al., 1979). Before taking action, it is imperative to accurately forecast the worldwide demand for diverse fuel sources, the allocation of carbon dioxide across the different reservoirs in the carbon cycle, the repercussions of carbon dioxide fluctuations on climate at both global and regional levels, as well as the sociopolitical and environmental ramifications that may arise from climate change. Additionally, it is crucial to devise strategies for curtailing the volume of carbon dioxide emissions generated by energy production and for impacting the allocation of carbon dioxide among the various reservoirs in the carbon cycle, with the ultimate goal of reducing atmospheric levels.

8.2. REGIONAL

The health implications of different coal technologies at a regional level encompass fatalities and incapacitations resulting from respiratory ailments linked to the inhalation of pollutants. These pollutants include emissions released into the atmosphere during combustion and the byproducts generated during the transportation of pollutants through the air (Gordon et al., 1979). Additionally, accidental fatalities and serious injuries are associated with coal mining and transportation.

Public health may be impacted by a third category which involves fatalities and incapacitations resulting from cancer and genetic harm linked to trace pollutants that could be discharged while burning synthetic liquids derived from coal (Zhang et al., 2010). More comprehensive data is required to evaluate the likelihood and magnitude of said contamination before determining its significance. Ongoing research is being conducted to compare the carcinogenic properties of artificial and natural crude oils. However, the current findings only allow for limited conclusions regarding the prediction of comparative risks. To obtain accurate assessments regarding the potential elevation of cancer risks associated with synthetic fuels, acquiring additional information about the constituents of forthcoming synthetic fuels, the release of minute pollutants, and the extent of exposure to such pollutants is imperative (Erten & Utlu, 2020). Furthermore, it is necessary to establish the correlation between the magnitude of exposure and the consequent effects on human health.

Did You Know?

The first coal miner disaster occurred in England in 1638 when a mine explosion killed 17 people.

8.2.1. Effects on Health From Combustion Products

The release of combustion byproducts stemming from diverse coal technologies is anticipated to primarily occur during the procedures of coal conversion and electricity generation, as well as during the ultimate utilization of the fuels (Avakian et al., 2002). The resultant emissions engender many transformation products predominantly generated through photochemical reactions in the atmosphere.

A vast array of transformation and combustion products, numbering in the hundreds, if not thousands, can elicit harmful health effects, either in isolation or through synergistic interactions with other compounds. This section presents a diverse array of projections regarding the potential impacts yet refrains from drawing any definitive inferences regarding these effects' likelihood or degree of occurrence. The potential impact's severity is contingent upon various factors, including the pollutant's characteristics (both chemical and physical), fluctuations in its concentration over time, and the conditions of exposure (such as duration, environmental factors like temperature and humidity, and the existence of other chemical species or physical agents that may influence the impact) (Wakefield, 2010). Additionally, the characteristics of the exposed population, such as their health status, nutritional state, and susceptibility (e.g., the elderly or young), also play a role.

There is a lack of data regarding the extent of exposure to various toxic pollutants. Limited information is available regarding the association between these substances and disease. Extensive research has been conducted on six pollutants' health impacts, primarily from fossil fuel combustion or atmospheric transformation processes. These pollutants include sulfur oxides, carbon monoxide, photochemical oxidants (including ozone), nitrogen dioxide, particulate matter, and non-methane hydrocarbons. The air pollutants subject to regulation in the United States are commonly known as criteria or controlled air pollutants. Preliminary National Ambient Air Quality Standards have been established based on extant data (Neumann, 2001). The validity of the foundation upon which the health criteria for controlled atmospheric contaminants are established is subject to scrutiny. While acknowledging its significance, the matter of resolution is deemed to be outside the purview of this particular chapter. Significant ambiguities exist regarding the potential health ramifications of various unregulated contaminants, including organic aerosols, acid sulfates, and nitrate particles (Sgro et al., 2012). Due to the lack of available health-related information about said pollutants, the ensuing discourse will employ the three criteria of contaminants and particulates as the basis for pollution indexing, despite their inadequacy as a measure for aerosol-type pollutants.

Research on the health impacts of air pollutants encompasses animal laboratory studies, occupational group investigations, and epidemiological investigations of the broader populace. The breadth of this section limits the possibility of conducting a comprehensive examination of all the significant research. The discourse is constrained to matters or contaminants of utmost importance (Ganhi & Lyon, 1998). The scope of references is confined to significant evaluations that assess and allude to numerous individual studies. The studies exhibit several limitations that constrain their applicability.

The primary challenge in generalizing findings from animal laboratory studies to humans is the physiological distinctions between humans and other species (Sarofim & Suk, 1994). Moreover, most animal studies are structured to examine brief, intense exposures, which presents additional challenges in generalizing the findings to prolonged, low-level exposures experienced by humans. Conducting long-term, low-dose experiments on animals is not only significantly costlier than short-term, high-dose experiments but is also constrained by the comparatively shorter life spans of laboratory animals compared to humans. Extrapolating from the fixed concentrations of individual pollutants to which animals are typically exposed to the varying levels of combined exposures to multiple pollutants characteristic of humans is a highly challenging task. The utilization of human subjects in laboratory studies has the potential to mitigate certain issues related to extrapolation (Boman et al., 2003). However, such an approach would give rise to ethical and practical dilemmas.

The generalizability of findings from studies conducted on occupational groups is constrained. There exist variations in the exposure conditions between occupational groups and the general population exposed to pollutants. These differences pertain to the composition of pollutants, their levels, the environmental parameters during exposure (such as temperature and humidity), and the time frame of exposure (typically 8 hours per day, 5 days per week) (Lomnicki et al., 2014). In addition, it should be noted that the demographic composition of the working population differs from that of the overall population, as it excludes vulnerable segments such as the elderly and young children. Consequently, the absence of an effect in a study of specific occupations does not necessarily imply the absence of such an effect in the broader population. If a demonstrated effect is observed, it can be anticipated that a larger effect will be evident in the broader population. Notwithstanding, these investigations hold significance as they can furnish insights into the quantities of pollutants ingested and expelled, and they ought to result in more accurate assessments of the implications for human beings of amalgamated pollutants present in the atmosphere.

Epidemiological investigations conducted on the general populace are subject to significant constraints. Typically, it is not feasible to establish a definitive causal relationship between exposure to a solitary pollutant and the manifestation of adverse health effects. Rather, it is only possible to suggest the existence of a correlation between the effects and a combination of pollutants. It is imperative to conduct experimental studies to assess the toxicity of the compound's constituent elements (Weidemann et al., 2016). However, the feasibility of such studies in the immediate future appears to be remote. The utilization of sulfur oxide levels as a substitute for the conglomeration of pollutants is anticipated to persist. Limitations exist in estimating the dosage, quantifying the effects, and accounting for environmental fluctuations and socioeconomic and racial disparities within the exposed population. Competing risks like smoking cigarettes can frequently lead to significant challenges in interpreting results. This is due to the possibility of such risks entirely obscuring the effects of chronic exposure to low levels of air pollution, posing serious problems (Geng et al., 2017). Incorporating data from all three types of studies can enhance the assessment of health impacts.

Given the present state of knowledge, it is not feasible to undertake an individualized evaluation of the potential health effects associated with each air pollutant that the populace may encounter. Furthermore, given that exposure to multiple pollutants concurrently can result in synergistic effects that deviate from the additive effects of individual pollutants, it is advisable to assess pollutants collectively. A proposed resolution to this issue involves utilizing a solitary pollutant as a comprehensive indicator for all other pollutants and assessing the potential health ramifications of this particular exposure. The emission rate of the index pollutant emanating from a specific energy plant can be determined by analyzing the attributes of the energy plant, such as the type of fuel utilized and the emission control apparatus implemented (Holleman et al., 1983). Subsequently, it is possible to create an air pollution model that incorporates the transportation of the designated pollutant, its atmospheric reactions, the exposure of specific populations to the pollutant, and the corresponding responses to the exposure. It is imperative to acknowledge that each presumption is accompanied by errors, which collectively result in significant uncertainties and wide-ranging impact estimates.

8.2.2. Impacts of Coal Transportation

The conveyance of coal represents a significant contributor to the regional health impacts. Approximately 66% of the transportation of coal in the United States is facilitated through train transportation.

Table 8.1 presents estimations of the incidence of accidental fatalities and injuries that occur during the transportation of coal (Mamurekli, 2010). The assessment of fatalities and injuries in railway incidents comprises two distinct elements: firstly, those resulting from collisions between trains and motor vehicles at rail grade crossings, and secondly, those arising from harm to rail passengers (Fatah, 2008). The incidence of unintended fatalities and injuries resulting from transportation mishaps is contingent upon various factors, including the quantity of grade-level crossings, automated safety mechanisms at junctions and on trains, and the upkeep of the rail bed. The attribution of fatalities and physical harm resulting from train mishaps has been assigned to coal rail conveyance through two distinct means. One metric is calculated based on the distance traveled in miles, while the other is based on the weight of cargo transported in ton-miles (Liu & Liu, 2020).

Annual fatality estimates for the transportation of coal to a 1,000-MW power plant, which consumes 1.9 million tons of coal per year, have been calculated based on train miles. It is important to note that the calculations only consider unit coal trains that carry 10,000 tons of coal each (Goswami, 2015). As mentioned earlier, the assumption can be categorized as conservative, given that mixed trains are responsible for conveying approximately 66.67% of the total coal transported via trains in contemporary times. Compared to unit trains, their coal-carrying capacity is significantly lower, resulting in increased transportation mileage for delivering millions of tons of coal.

Table 8.1. Rates of Fatalities and Injuries Resulting from Various Transportation Modes

Transportation Mode	Fatalities (cases/)	Injuries (cases/)
Rail	0.062–0.075	0.60–1.00
Trucking	0.032	0.1
Barge	0.0019	0.0032
Slurry pipeline	0.0019	0.0032

Source: C. P. Bliss, Creative Commons License.

The annual fatality count varies across different power plants, with the lowest rate of 0.11 being observed for a plant in the Eastern Interior Region fueled by coal sourced from southern Illinois, requiring a round-trip distance of 200 miles (Wenlong et al., 2023). In contrast, the highest fatality rate of 1.45 is associated with plants in the Eastern Interior Region that rely on coal from Wyoming, necessitating a round-trip distance of 1,900 miles. Assuming a mean distance of 600 miles per round trip and considering the proportion of general public deaths as a subset of all accidental deaths, the resulting annual fatality rate is 0.15 per million tons of coal.

Other scholars have posited that the computation of unintentional collisions is predicated upon the metric of ton-miles. The authors' calculations of accidental deaths per million tons of coal for the general public, based on an assumed energy content of 25 x 10 Btu per ton of coal, yielded estimates of approximately 1.5 and 15. These values are roughly 10 times higher than the corresponding train-miles estimates (Mishra, 2004).

8.3. LOCAL

According to the present understanding, the primary local health and safety ramifications are linked to coal mining. The aforementioned coal technologies are responsible for a significant proportion of occupational fatalities, injuries, and adverse health effects.

There exist various local ramifications for laborers in coal technologies. Based on current knowledge, the most significant factors contributing to accidental deaths and injuries in the context of energy conversion facilities are, in order of importance, the construction and operation of such facilities, the transportation of coal, and the preparation and storage of coal. Insufficient implementation of controls in coal transformation plants may result in heightened susceptibility of workers to mortality and incapacitation due to cancer (Mohring, 1997). Implementing controls would encompass various aspects such as engineering factors, workplace sanitation evaluation, medical monitoring, personal sanitation, workplace procedures, and worker health education.

8.3.1. Mining

8.3.1.1. Effects on Health

The U.S. Social Security Administration uses chronic bronchitis, CWP, and obstructive airway diseases to determine black lung disease patients' eligibility for social security benefits. Coal miners had three times more chronic bronchitis and obstructive airway disorders than CWP (Castleden et al., 2011). However, CWP is a clinically diagnosable respiratory condition connected to persistent coal dust inhalation. Thus, it better measures coal mining's respiratory effects.

Coal dust exposure increases the risk of CWP. This occurs most often among underground coal miners who work near the coal seam's frontline, where dust particles are most concentrated. Chronic widespread pain (CWP) is rarest in surface and underground coal miners (Von der Goltz & Barnwal, 2019). Thus, regulating respirable coal dust in mining contexts is the best way to reduce new Coal Workers' Pneumoconiosis instances. The British experience of limiting coal dust to 2 mg/m^3 supports the above claim. At the prescribed concentrations, British underground workers' CWP rate has dropped to 3%, and these workers show very minimal symptoms. Local environmental variables, geographical differences, the length between exposure and diagnosis, and the severity of CWP affect its prevalence and type. Dissimilarities in the chemical composition and magnitude distribution of coal particulate matter formed from different coal layers may explain the observed differences.

CWP prevalence in the US varies widely. According to the National Institute of Occupational Safety and Health, 10–15% of underground coal miners have the disease to varied degrees. The National Academy of Sciences Committee on Mineral Resources and the Environment believes that 30% of underground miners have the condition. In 1976, surface mining had 4.0 occurrences per million tons. The Federal Coal Mine Health and Safety Act of 1969 and the Black Lung Benefits Act of 1972 compensate US miners incapacitated by CWP. From 1970 through 1973, miners and their families received $1.7 billion in compensation, rising to over $1 billion 1973. The 1972 statute forbade denying compensation to miners based on a regular roentgenogram, which caused costs to rise quickly. To reduce CWP, the US Coal Mine Health and Safety Acts of 1969 and 1977 limited respirable dust in coal mines (Zhang et al., 2012). CWP prevalence predictions are questionable. The Safety Act's respirable dust limits and the move from underground to surface coal mining would significantly reduce the number of cases per million tons of coal extracted.

Coal Mining Accidents: Coal extraction is the main source of unintended deaths and injuries to workers in coal-based energy infrastructure. Underground coal mining is less safe. In 1975, underground mining had

0.48 accidents and 37.11 fatalities per million person-hours, whereas surface mining had 0.35 and 18.10. Surface coal mining has fewer accidents per million tons extracted due to its increased productivity and other variables. Surface coal mining has one-fourth of the unintentional deaths and one-sixth of the injuries per million tons extracted (Petavratzi et al., 2005). In 1975, coal mining caused 200 accidental deaths, most of the underground workers. Over the past 50 years, underground coal mining fatalities and injuries per million tons have decreased. Productivity tripled between 1949 and 1969, reducing accidental rates per million tons of coal. Safety improvements caused the 1970–1976 decline. From 1970 to 1976, output per person-day dropped 40%. Mine safety improved by 60%.

8.3.2. Construction of Facilities

Construction fatalities and injuries significantly affect workers' professional health and safety within the local context. The construction industry experiences a comparatively lower impact than the mining industry, with 7.4 fatalities per million person-years and 30,000 nonfatal injuries resulting in disabilities per million person-years (Li & Jia, 2017). Given that the primary workforce demands pertain to the erection of energy conversion infrastructure, the preponderance of fatalities and physical harm is linked to this particular stage of the coal process. The coal systems necessitate significant supplementary workforce needs, particularly for establishing mining infrastructure. According to the coal requirements over the lifespan of the energy facilities, their workforce constitutes a proportion of manpower needs below 5% (Triegel, 1980).

8.3.3. Carbonization

8.3.3.1. Occupational Hazards

Cancer: Cancer has emerged as the predominant health issue among employees working in carbonization facilities (Barak, 2020). The concern is predicated upon two distinct categories of research:

i. Epidemiological investigations have revealed an elevated likelihood of cancer among individuals regularly exposed to coal-derived substances, including those employed in coal gas, coking, coal hydrogenation industries, street pavers, and asphalt workers.

ii. Laboratory investigations have been conducted to ascertain the potential carcinogenicity of diverse coal products and their constituent chemicals, employing animal models, bacterial cultures, and human cell lines.

Redmond endeavored to establish a correlation between the risk of lung cancer in a Coke plant and factors such as work area and length of employment, as evidenced by Table 8.2 (He et al., 2013).

The research encompassed approximately 60,000 steel industry employees across seven distinct facilities. The study revealed that male individuals who were employed for five years or more in locations where they were exposed to coke-oven emissions at the highest level, specifically at the top of the ovens, had an approximately 90 £m increase in lung cancer mortality compared to their counterparts who worked in other areas of the steel industry (Fakkaew et al., 2018). As anticipated, the surplus mortality rates due to cancer were significantly reduced or insignificant among coke industry employees working in cleaner areas and with shorter duration of exposure.

Lloyd conducted a study wherein the relative risk of mortality due to various causes was determined among 2048 full-time coke-oven workers based on their race and workstation within the plant. The research spanned a period of eight years, from 1953 to 1961. Lloyd's research revealed that the mortality rate of malignant neoplasms of the lung among coke-oven workers was 2.5 times higher than the anticipated rate (Dananjayan et al., 2016).

Table 8.2. Observed and Expected Deaths and Relative Risks for Malignant Neoplasms for Coke-Oven Workers

Cause of Death	**Total Oven**			**5 Years or More, Coke Oven**		
	Observed Deaths	**Expected Deaths**	**Relative Risk**	**Observed Deaths**	**Expected Deaths**	**Relative Risk**
Malignant neoplasms, all sites.	145	122.1	1.34	114	82.6	1.69
Bronchus, lung, and trachea	69	41.5	2.85	55	28.0	3.48
Other respiratory system	1	2.4	*d*	1	1.8	*d*
Buccal and pharynx	3	3.4	*d*	1	2.2	*d*

Digestive organs and peritoneum	31	37.0	0.77	25	26.0	0.95
Prostate	12	8.8	1.64	11	7.4	1.83
Other genital organs	0	0.4	*d*	0	0.3	*d*
Kidney	8	2.6	7.49	5	1.6	5.69
Bladder and other urinary organs	1	1.5	*d*	1	1.	*d*
Skin	1	1.5	*d*	1	0.8	*d*
Leukemias and lymphomas	9	10.8	0.76	5	6.0	0.77
Other and unspecified sites	10	12.1	0.69	9	7.3	1.46

Source: Martin Anderson, Creative Commons License.

The study found that the mortality risk for full topside workers was 6.5 times higher than that for men employed solely at the side of the oven, as indicated by the standard mortality ratio of 818 versus 125 (Huntzinger et al., 2009). No significant difference was observed in the mortality rate of side oven workers from other factors compared to the expected rate.

Mazumdar successfully established a correlation between the exposure of coke workers to coal-tar pitch volatiles (CTPV) and the heightened risk of cancer, as evidenced in Table 8.3 (Sun et al., 2012). The researchers utilized Redmond's risk factors with exposure data from the Pennsylvania Department of Health, specifically from 10 coke-oven installations. The data presented significant variations in the level of exposure to coke-oven emissions among the different job positions. The range of exposures observed in the study varied from 0.50 mg/m° for the low average to 3.23 mg/m° for the exceptionally high average exposure. The findings suggest a significant correlation between the duration and quantity of exposure and the incidence of cancer, specifically lung cancer. The duration between initial exposure to CTPV and mortality was 10 to 40 years, with an average of 25 years. There is no evidence of increased cancer incidence among workers exposed to the established threshold for an average of 30 years (Kershaw & Smart, 1994).

Ø **Identification of Carcinogens:** Bituminous coal typically comprises 14–5% volatile material on a dry basis, while anthracite coal typically contains 2–14% volatile material on a dry basis (Doll et al., 1965). The category of volatile matter encompasses a multitude of organic compounds, some of which possess the potential to induce carcinogenic effects. This group includes aromatic in nature and heterocyclic compounds.

Table 8.3. Observed and Expected Deaths and Relative Risks for Malignant Neoplasms for Coke-Oven Workers Exposed to Less Than Five Years

Cause of Death	**< 5 Years, Coke Oven**		
	Observed Deaths	**Expected Deaths**	**Relative Risk**
Malignant neoplasms, all sites	31	38.6	0.74
Bronchus, lung, and trachea	14	9.9	1.70
Other respiratory system	0	1.0	*d*
Buccal and pharynx	2	1.4	*d*
Digestive organs and peritoneum	6	12.2	0.42
Prostate	1	1.4	*d*
Other genital organs	0	0.2	*d*
Kidney	3	0.7	*d*
Bladder and other urinary organs	0	0.4	*d*
Skin	0	0.8	*d*
Leukemias and lymphomas	4	5.3	0.68
Other and unspecified sites	1	5.2	*d*

Source: Martin Anderson, Creative Commons License.

PAHs are considered to be the most significant potential carcinogens. Additional significant categories include diverse classes of organic compounds, such as polycyclic azaheterocyclic compounds, aromatic compounds, and benzenes, alongside inorganic trace elements. Coal tars generated through high-temperature processes exhibit significantly greater carcinogenicity than those observed among laborers at coal gasification plants in England (Borm, 1997). The study encompasses a cohort of 11,500 individuals who were regularly or intermittently engaged in the carbonization process of gas plants for a duration exceeding five years.

The cohort of workers involved in carbonization, subjected to higher levels of coal and its byproducts, exhibited a standardized annual mortality rate of 2.86 for bronchitis, in contrast to the rate of 1.36 observed in those exposed to lower temperatures. The destructive distillation of coal at a temperature below 450°C (842°F) produces minimal quantities of carcinogenic substances (Skodras et al., 2006). The production of substances with carcinogenic properties experiences a significant surge within the temperature range of 450 to 550°C (842—958°F) associated with coking, and subsequently, the rate of increase persists albeit at a slower pace until the temperature reaches 1,250°C (2,282°F). In 1973, an endeavor was undertaken to establish a correlation between cooking temperature and the prevalence of lung cancer among operators (Table 8.4) (Li et al., 2019).

Table 8.4. Observed and Expected Deaths and Relative Risks for Malignant Neoplasms for Selected Causes of Deaths for a Coke-Oven Worker: 1951–1966

Cause of Death	**5 Years or More, Coke Oven**			**<5 Year, Coke Oven**		
	Observed Deaths	**Expected Deaths**	**Relative Risk**	**Observed Deaths**	**Expected Deaths**	**Relative Risk**
All causes	383	365.9	1.08	206	235.2	0.83
Malignant neoplasms-lung, bronchus, and trachea.	55	28	3.48	14	9.9	1.70
Malignant neoplasms-genitourinary organs.	17	10.6	*2.01*	4	2.6	*e*
Other malignant neoplasms.	42	44.0	*0.93*	13	26.0	0.40
Tuberculosis of the respiratory system.	5	5.0	1.00	3	2.9	*e*

Other diseases of the respiratory system.	20	16.8	1.20	10	11.0	0.88
Cardiovascular and renal diseases.	173	186.2	*0.90*	81	96.4	0.79
Accidents	20	21.9	0.88	42	38.6	1.12
All other causes	51	53.4	*0.94*	39	47.6	0.77

Source: C. K. Redmond, Creative Commons License.

Other Health Impairments. Data on increased mortality from all causes for non-gas-plant workers in England and Wales are available. These workers also exhibited a death rate 1.19 for other respiratory diseases versus 1.14 for the general population. For the 10 groups of diseases considered, this occupational group also experienced the following increased death rates compared to those for the general public: cancer of the lung, 3.06 versus 1.81 for the general population; cancer of the bladder, 0.31 versus 0.14; cancer of skin and scrotum, 0.07 versus 0.02; all causes, 17.21 versus 16.35 (Maqhuzu et al., 2019). The data are similar to those discussed in the preceding section. It is interesting to note that the death rates of this occupational group were lower than those of non-gas-plant workers in England and Wales for the following causes: pneumoconiosis, 0.05 versus 0.07; arteriosclerotic and other heart disease, 3.34 versus 4.48; other cancers, 1.96 versus 2.29; other causes, 3.59 versus 4.31. In the case of accidental deaths, this group had a slightly higher rate: 0.76 versus 0.72.

The occupational group with lighter or intermittent exposure to coal products showed the same death rates from respiratory diseases as the general population.

The topic under consideration is related to health impairments. Data on elevated all-cause mortality rates among non-gas-plant workers in England and Wales exists. The workers in question demonstrated a mortality rate of 1.19 for other respiratory ailments, as opposed to 1.14 for the wider populace. Concerning the 10 categories of illnesses under consideration, this particular professional cohort exhibited elevated mortality rates in comparison to the wider populace for the following conditions: The incidence of lung cancer among individuals with a history of smoking is 3.06, which is higher than the rate of 1.81 observed in the general population (Cheng et al., 2022). Similarly, the incidence of bladder cancer is 0.31 compared to 0.14 in the general population, while skin and scrotum cancer incidence is 0.07 compared to 0.02. The overall incidence of all causes of cancer is 17.21 in individuals with a history of smoking, which is slightly higher than the rate of 16.35 observed in the general population. The aforementioned data bear a resemblance to the ones that were deliberated in the section that followed. Notably, the mortality rates for the specified causes among gas plant workers in England and Wales were comparatively lower than those of their non-gas plant counterparts. Specifically, the rates for pneumoconiosis were 0.05 as opposed to 0.07. For arteriosclerotic and other heart diseases, it was 3.34 versus 4.48; for other cancers, it was 1.96 versus 2.29; and for other causes, it was. This group exhibited a marginally elevated rate in the context of unintentional fatalities, with figures of 0.76 and 0.72, respectively (Mugoronji et al., 2022).

8.3.4. Combustion

The available data on morbidity and mortality between workers in coal-burning electric power plants and their contact with gaseous or aerosol emissions from coal combustion are limited. The literature currently lacks statistically significant measurements about heightened mortality or morbidity. At times, measurements are taken to monitor the levels of combustion byproducts and hydrocarbons. Certain measurements have indicated intermittent peak values that surpass the threshold limit values suggested by the National Institute for Occupational Safety and Health. However, a contentious debate exists regarding whether these exposures are adequate to cause any detrimental effects on health.

8.3.4.1. Organic Contaminants

Polynuclear aromatic hydrocarbons and polynuclear azaheterocyclic compounds are the primary organic contaminants that are of concern and are generated during coal combustion (Sun et al., 2018). The release of these pollutants into the environment can occur through various means, such as the discharge into the atmosphere during routine coal combustion procedures, the drainage and infiltration from insufficient waste management and storage practices and using material waste by-products for purposes such as soil conditioning and landfill.

As mentioned earlier, the management of contamination stemming from any of the sources can be achieved through implementing waste containment measures, utilizing control technologies to mitigate environmental pollution, and imposing restrictions on the utilization of polluted products and by-products (Rout et al., 2019) (Table 8.5).

Table 8.5. Number Employed Observed and Expected Deaths and Standardized Mortality Ratios

Work Area and Race	Number Employed Observed Deaths	All Causes			Malignant Neoplasms of Lung		
		Expected Deaths	SMR	Observed Deaths	Expected Deaths	SMR	
Total coke oven	2,048	184	173.1	106	31	12.3	252
White	993	80	80.2	100	8	4.7	170
Nonwhite	1,055	104	92.9	112	23	7.6	303
Side oven	1,431	106	114.6	92	10	8.0	125
White	606	45	47.2	95	5	2.7	185
Nonwhite	825	61	67.3	91	5	5.4	93
Partial topside	315	24	26.3	91	2	1.7	–
White	307	22	25.1	88	2	1.6	–
Nonwhite	8	2	1.2	–	0	0.1	–
Full topside	302	2	1.2	168	19	2.6	731
White	80	13	7.8	167	1	0.5	–
Nonwhite	222	41	24.4	168	18	2.2	818

Other Malignant Neoplasms			Heart Disease			All Other Causes		
Observed Deaths	Expected Deaths	SMR	Observed Deaths	Expected Deaths	SMR	Observed Deaths	Expected Deaths	SMR
27	26.9	100	49	61.5	80	77	72.4	106
12	12.2	98	29	33.9	86	31	29.4	105
15	14.8	101	20	27.6	72	46	43.0	107
13	17.5	74	34	40.0	85	49	49.1	100
6	7.3	82	18	19.9	90	16	17.4	92
7	10.3	68	16	20.1	80	33	31.6	104
6	4.0	150	5	10.9	46	11	9.7	113
5	3.8	132	5	10.5	48	10	9.2	109
1	0.2	–	0	0.3	–	1	0.5	–
8	5.4	148	10	10.6	94	17	13.6	125

1	1.2	–	6	3.5	171	5	2.8	179
7	4.3	163	4	7.2	56	12	10.8	111

Source: J. W. Lloyd, Creative Commons License.

The current understanding of the potential health impacts of trace organic contaminants is inadequate for quantitative or comprehensive qualitative evaluation. The available information on potential air emissions and subsequent human exposure to contaminants from coal systems is limited. To date, there is a shortage of literature regarding potential exposures linked to the contamination of water supplies and agricultural products resulting from runoff and seepage waters. The assessment of health effects, specifically carcinogenicity, is complex due to various factors such as exposure levels to trace contaminants, interactions among carcinogens, co-carcinogens, carcinogen initiators, and other organic agents and components (Luthy et al., 1997). Insufficient characterization of emitted pollutants hinders the estimation of potential health impacts. Additional investigation is required to comprehend the mechanisms of transformation and the potential impacts of transformation products on vulnerable organisms. Nonetheless, it is possible to generate preliminary evaluations of the health consequences that may arise from the discharge of diverse polynuclear aromatic hydrocarbons during coal combustion procedures by utilizing BaP or benzene-soluble organics as substitutes (Table 8.6).

Table 8.6. Temperature Range of Carbonization and Excess of Lung Cancers Reported

Type of Carbonizing Chamber	**Temperature Range (°C)**	**Excess of Lung Cancers Reported (%)**
Vertical retorts	400–500	27
Horizontal retorts	900–1,100	83
Coke ovens	1,200–1,400	255
Japanese gas generators	1,500	800

Source: Martin Anderson, Creative Commons License.

Information pertaining to the emissions and atmospheric concentrations of BaP and benzene-soluble organics is readily accessible. Notably, strong correlations exist between their atmospheric concentrations and those of diverse polynuclear aromatic compounds (Basha et al., 2009).

The primary origins of BaP discharges are domestic coal and wood furnaces, domestic wood fireplaces, minor housing and industrial boilers, automotive emissions, and diverse industrial procedures, such as coke manufacturing. In 1972, the estimated BaP emissions from coal-burning power plants in the United States were less than one ton, which accounted for approximately 0.33% of the estimated national BaP emission. According to the coke-oven estimates, the annual surplus cancer risk linked with a coal-burning electric power plant of 1,000-MWe capacity is approximately 0.005 (Dempster et al., 1997).

8.3.4.2. Inorganic Contaminants

Similar conclusions can be drawn for organic constituents regarding the potential health effects of exposure to inorganic contaminants that arise from coal combustion. The primary origin of discharges of inorganic substances from coal is attributed to its mineral composition, which predominantly comprises silicon, calcium, aluminum, and iron, alongside a plethora of other elements, albeit in relatively lower concentrations (as depicted in Table 8.7) (Raeva et al., 2011). Certain inorganic components present in coal possess the potential to induce carcinogenicity. The aforementioned elements encompass metallic and nonmetallic categories, namely arsenic, beryllium, cadmium, cobalt, chromium, nickel, lead, and selenium.

Table 8.7. Minor and Trace Elements in Coal and Earth's Crust

Element	Concentration (ppm)		
	Coal (mean)	**Coal (range)**	**Earth's Crust (mean)**
Arsenic	14.0	0.5–93.0	1.8
Beryllium	1.6	0.2–4.0	2.8
Cadmium	2.5	0.1–65.0	0.2
Chromium	13.8	4.0–54.0	100.0
Cobalt	9.6	1.0–43.0	25.0
Iron	1.9	0.3–4.3	5600.0
Lead	34.8	4.0–218.0	12.5
Nickel	21.1	3.0–80.0	75.0
Selenium	2.1	0.5–7.7	0.05

Source: M. R. Kornreich, Creative Commons License.

The significant variations in trace element concentrations observed across coal samples are noteworthy. Differences in the chemical makeup can be observed even among coal specimens obtained from identical seams (Jelonek et al., 2021). The projection of potential emissions of trace elements from a coal-utilizing plant and the subsequent impacts on health can be subject to significant errors.

8.3.5. Conversion

Sexton issued a coal conversion plant cancer report. Institute, West Virginia, has a coal liquefaction pilot facility. From 1952 through the mid-1950s, 259 workers produced 300 tons of coal hydrogenation products each day at the factory. Industrial hygiene was implemented due to cancer dangers. It taught employees to apply barrier cream, protective clothes, frequent clothing changes, and skin examinations. Sexton found 42 precancerous skin lesions between 1954 and 1959. This is 16–37 times West Virginia and US rates (Tchapda & Pisupati, 2014).

Since some of the guys had been exposed to identical chemicals in lab work, their chemical exposure ranged from months to years. Despite cleaning and protective precautions, spot inspections on factory workers showed skin contamination, calling the industrial hygiene program into doubt (Tanaka et al., 2002). These results only suggest that coal liquefaction products can cause skin cancer. Exposure has not been linked to skin cancer or other cancers. Three basal cell carcinomas were reported at Exxon Research and Engineering after 966 person-years of coal hydrogenation in liquid products. The list of carcinogens in coal conversion process streams shows the range of potential carcinogenic pollutants (Klasson et al., 1993). However, more information on their relative hazards, emission rates, and projected exposures is needed to assess their health effects.

8.4. SUMMARY

Coal utilization particularly in power generation and industrial processes has significant impacts on health and safety. The burning of coal releases various pollutants into the air and these pollutants contribute to the formation of smog, acid rain, and ground-level ozone. Coal combustion releases fine particulate matter known as PM2.5, which can penetrate deep into the lungs. Coal is a significant contributor to greenhouse gas emissions, particularly carbon dioxide which is a major driver of climate change. Workers in coal mines, power plants, and other coal-related industries face various occupational hazards such as physical injuries and exposure to toxic gases. The overall impacts on health and safety underscore the need for transitioning to cleaner and more sustainable energy alternatives.

REFERENCES

1. Avakian, M. D., Dellinger, B., Fiedler, H., Gullet, B., Koshland, C., Marklund, S., & Suk, W. A., (2002). The origin, fate, and health effects of combustion by-products: A research framework. *Environmental Health Perspectives, 110*(11), 1155–1162.
2. Balat, M., (2007). Influence of coal as an energy source on environmental pollution. *Energy Sources, Part A: Recovery, Utilization, and Environmental Effects, 29*(7), 581–589.
3. Barak, O., (2020). *Powering Empire: How Coal Made the Middle East and Sparked Global Carbonization?* (Vol. 3, No. 2, pp. 8–9). University of California Press.
4. Basha, C. A., Chithra, E., & Sripriyalakshmi, N. K., (2009). Electro-degradation and biological oxidation of non-biodegradable organic contaminants. *Chemical Engineering Journal, 149*(1–3), 25–34.
5. Bezdek, R. H., (1993). The environmental, health, and safety implications of solar energy in central station power production. *Energy, 18*(6), 681–685.
6. Boman, B. C., Forsberg, A. B., & Järvholm, B. G., (2003). Adverse health effects from ambient air pollution in relation to residential wood combustion in modern society. *Scandinavian Journal of Work, Environment, and Health, 1*, 251–260.
7. Borm, P. J., (1997). Toxicity and occupational health hazards of coal fly ash (CFA). A review of data and comparison to coal mine dust. *The Annals of Occupational Hygiene, 41*(6), 659–676.
8. Bukhari, S. S., Behin, J., Kazemian, H., & Rohani, S., (2015). Conversion of coal fly ash to zeolite utilizing microwave and ultrasound energies: A review. *Fuel, 140*(1), 250–266.
9. Castleden, W. M., Shearman, D., Crisp, G., & Finch, P., (2011). The mining and burning of coal: Effects on health and the environment. *The Medical Journal of Australia, 195*(6), 333–335.
10. Cheng, L., Wei, W., Guo, A., Zhang, C., Sha, K., Wang, R., & Cheng, S., (2022). Health risk assessment of hazardous VOCs and its associations with exposure duration and protection measures for coking industry workers. *Journal of Cleaner Production, 379*(1), 134–919.
11. Dananjayan, R. R. T., Kandasamy, P., & Andimuthu, R., (2016). Direct mineral carbonation of coal fly ash for CO_2 sequestration. *Journal of Cleaner Production, 112*(1), 4173–4182.

12. Dempster, H. S., Sherwood Lollar, B., & Feenstra, S., (1997). Tracing organic contaminants in groundwater: a new methodology using compound-specific isotopic analysis. *Environmental Science and Technology, 31*(11), 3193–3197.
13. Doll, R., Fisher, R. E. W., Gammon, E. J., Gunn, W., Hughes, G. O., Tyrer, F. H., & Wilson, W., (1965). Mortality of gas workers with special reference to cancers of the lung and bladder, chronic bronchitis, and pneumoconiosis. *Occupational and Environmental Medicine, 22*(1), 1–12.
14. Erten, B., & Utlu, Z., (2020). Photovoltaic system configurations: An occupational health and safety assessment. *Greenhouse Gases: Science and Technology, 10*(4), 809–828.
15. Fakkaew, K., Koottatep, T., & Polprasert, C., (2018). Fecal sludge treatment and utilization by hydrothermal carbonization. *Journal of Environmental Management, 216*(1), 421–426.
16. Fatah, L., (2008). *The Impacts of Coal Mining on the Economy and Environment of South Kalimantan Province, Indonesia* (Vol. 1, pp. 85–98). ASEAN economic bulletin, Tanaka, Tanaka.
17. Finkelman, R. B., Orem, W., Castranova, V., Tatu, C. A., Belkin, H. E., Zheng, B., & Bates, A. L., (2002). Health impacts of coal and coal use: Possible solutions. *International Journal of Coal Geology, 50*(1–4), 425–443.
18. Ganhi, S., & Lyon, R. E., (1998). Health hazards of combustion products from aircraft composite materials. *Federal Aviation Administration Technical Center Atlantic City NJ, 5*(1), 7–9.
19. Geng, P., Cao, E., Tan, Q., & Wei, L., (2017). Effects of alternative fuels on the combustion characteristics and emission products from diesel engines: A review. *Renewable and Sustainable Energy Reviews, 71*(1), 523–534.
20. Gitari, W. M., Petrik, L. F., Etchebers, O., Key, D. L., & Okujeni, C., (2008). Utilization of fly ash for treatment of coal mines wastewater: Solubility controls on major inorganic contaminants. *Fuel, 87*(12), 2450–2462.
21. Gordon, R. L., Manula, C. B., & Ramani, R. V., (1979). *Regulation of the Coal Industry-Health and Safety, Surface Mining and Other Issues* (Vol. 2, No. 1, pp. 6–8). Parts I and II (No. PB-80-176597). Pennsylvania State Univ., University Park (USA). Department of Mineral Economics.
22. Goswami, S., (2015). Impact of coal mining on environment. *European Researcher, 3*, 185–196.
23. He, L., Yu, D., Lv, W., Wu, J., & Xu, M., (2013). A novel method for CO_2 sequestration via indirect carbonation of coal fly ash. *Industrial and*

Engineering Chemistry Research, 52(43), 15138–15145.

24. Hendryx, M., Zullig, K. J., & Luo, J., (2020). Impacts of coal use on health. *Annual Review of Public Health, 41*(1), 397–415.
25. Holleman, J. W., Ross, R. H., & Carroll, J. W., (1983). *Problem Definition Study on the Health Effects of Diethyleneglycol Dinitrate, Triethylenegycol Dinitrate, and Trimethylolethane Trinitrate and Their Respective Combustion Products* (Vol. 1, pp. 7–10). Oak Ridge National Lab TN.
26. Huntzinger, D. N., Gierke, J. S., Sutter, L. L., Kawatra, S. K., & Eisele, T. C., (2009). Mineral carbonation for carbon sequestration in cement kiln dust from waste piles. *Journal of Hazardous Materials, 168*(1), 31–37.
27. Jelonek, Z., Fabiańska, M., & Jelonek, I., (2021). Quantitative assessment of organic and inorganic contaminants in charcoal. *Resources, 10*(7), 69.
28. Kershaw, J. R., & Smart, P. J., (1994). Extraction of coal-tar pitch and the effect on carbonization. *Carbon, 32*(1), 85–92.
29. Klasson, K. T., Ackerson, M. D., Clausen, E. C., & Gaddy, J. L., (1993). Biological conversion of coal and coal-derived synthesis gas. *Fuel, 72*(12), 1673–1678.
30. Li, Q., Qi, J., Jiang, J., Wu, J., Duan, L., Wang, S., & Hao, J., (2019). Significant reduction in air pollutant emissions from household cooking stoves by replacing raw solid fuels with their carbonized products. *Science of the Total Environment, 650*(1), 653–660.
31. Li, W., & Jia, X., (2017). Ground control issues on photovoltaic power generation facilities construction in coal sinkhole region. *Procedia Engineering, 191*(1), 98–103.
32. Liu, T., & Liu, S., (2020). The impacts of coal dust on miners' health: A review. *Environmental Research, 190*(1), 109–849.
33. Lomnicki, S., Gullett, B., Stöger, T., Kennedy, I., Diaz, J., Dugas, T. R., & Cormier, S. A., (2014). Combustion by-products and their health effects—combustion engineering and global health in the 21st century: Issues and challenges. *International Journal of Toxicology, 33*(1), 3–13.
34. Luthy, R. G., Aiken, G. R., Brusseau, M. L., Cunningham, S. D., Gschwend, P. M., Pignatello, J. J., & Westall, J. C., (1997). Sequestration of hydrophobic organic contaminants by geosorbents. *Environmental Science and Technology, 31*(12), 3341–3347.
35. Mamurekli, D., (2010). Environmental impacts of coal mining and coal utilization in the UK. *Acta Montanistica Slovaca, 15*(2), 134.
36. Maqhuzu, A. B., Yoshikawa, K., & Takahashi, F., (2019). The effect of coal alternative fuel from municipal solid wastes employing hydrothermal

carbonization on atmospheric pollutant emissions in Zimbabwe. *Science of the Total Environment, 668*(1), 743–759.

37. Mishra, U. C., (2004). Environmental impact of coal industry and thermal power plants in India. *Journal of Environmental Radioactivity, 72*(1–2), 35–40.
38. Mohring, R. P., (1997). Presidential address-The South African coal industry: Current position and future challenges. *Journal of the Southern African Institute of Mining and Metallurgy, 97*(5), 205–209.
39. Montes-Hernandez, G., Pérez-López, R., Renard, F., Nieto, J. M., & Charlet, L., (2009). Mineral sequestration of CO_2 by aqueous carbonation of coal combustion fly-ash. *Journal of hazardous Materials, 161*(2–3), 1347–1354.
40. Morris, S. C., Moskowitz, P. D., Sevian, W. A., Silberstein, S., & Hamilton, L. D., (1979). Coal conversion technologies: Some health and environmental effects. *Science, 206*(4419), 654–662.
41. Mugoronji, M., Manyuchi, M. M., Sukdeo, N., & Stinner, W., (2022). Techno-economic assessment for bio-coal production from brewers spent grain. *South African Journal of Chemical Engineering, 40*(1), 1–9.
42. Neumann, H. G., (2001). Health risk of combustion products: Toxicological considerations. *Chemosphere, 42*(5–7), 473–479.
43. Petavratzi, E., Kingman, S., & Lowndes, I., (2005). Particulates from mining operations: A review of sources, effects, and regulations. *Minerals Engineering, 18*(12), 1183–1199.
44. Raeva, A. A., Pierce, D. T., Seames, W. S., & Kozliak, E. I., (2011). A method for measuring the kinetics of organically associated inorganic contaminant vaporization during coal combustion. *Fuel Processing Technology, 92*(7), 1333–1339.
45. Ren, X. Y., Feng, X. B., Cao, J. P., Tang, W., Wang, Z. H., Yang, Z., & Zhao, X. Y., (2020). Catalytic conversion of coal and biomass volatiles: A review. *Energy and Fuels, 34*(9), 10307–10363.
46. Rout, L., Kumar, A., Achary, L. S. K., Barik, B., & Dash, P., (2019). Ionic liquid assisted combustion synthesis of ZnO and its modification by AuSn bimetallic nanoparticles: An efficient photocatalyst for degradation of organic contaminants. *Materials Chemistry and Physics, 232*(1), 339–353.
47. Sarofim, A. F., & Suk, W. A., (1994). Health effects of combustion by-products. *Environmental Health Perspectives, 102*(1), 237–244.
48. Sgro, L. A., D'Anna, A., & Minutolo, P., (2012). On the characterization of nanoparticles emitted from combustion sources related to understanding

their effects on health and climate. *Journal of Hazardous Materials, 211*(1), 420–426.

49. Skodras, G. E. O. R. G. I. O. S., Natas, P., Basinas, P., & Sakellaropoulos, G. P., (2006). Effects of pyrolysis temperature, residence time on the reactivity of clean coals produced from poor quality coals. *Global NEST Journal, 8*(2), 89–94.
50. Sun, J., Pan, L., Tsang, D. C., Zhan, Y., Zhu, L., & Li, X., (2018). Organic contamination and remediation in the agricultural soils of China: A critical review. *Science of the Total Environment, 615*(1), 724–740.
51. Sun, Y., Parikh, V., & Zhang, L., (2012). Sequestration of carbon dioxide by indirect mineralization using Victorian brown coal fly ash. *Journal of Hazardous Materials, 209*(1), 458–466.
52. Tanaka, H., Sakai, Y., & Hino, R., (2002). Formation of Na-A and-X zeolites from waste solutions in conversion of coal fly ash to zeolites. *Materials Research Bulletin, 37*(11), 1873–1884.
53. Tchapda, A. H., & Pisupati, S. V., (2014). A review of thermal co-conversion of coal and biomass/waste. *Energies, 7*(3), 1098–1148.
54. Triegel, E. K., (1980). *Attenuation by Soils of Selected Potential Contaminants from Coal Conversion Facilities: A Literature Review* (Vol. 1, pp. 6–9).
55. Vamvuka, D., (1999). Environmental impacts of coal utilization on the ecosystem. *Energy Exploration and Exploitation, 17*(6), 583–605.
56. Von der Goltz, J., & Barnwal, P., (2019). Mines: The local wealth and health effects of mineral mining in developing countries. *Journal of Development Economics, 139*(1), 1–16.
57. Wakefield, J. C., (2010). A toxicological review of the products of combustion. *Health Protection Agency, Center for Radiation, Chemical, and Environmental Hazards, Chemical Hazards and Poisons Division, 1*, pp. 5–7.
58. Weidemann, E., Andersson, P. L., Bidleman, T., Boman, C., Carlin, D. J., Collina, E., & Jansson, S., (2016). 14th congress of combustion by-products and their health effects – origin, fate, and health effects of combustion-related air pollutants in the coming era of bio-based energy sources. *Environmental Science and Pollution Research, 23*(1), 8141–8159.
59. Wenlong, Z., Nawaz, M. A., Sibghatullah, A., Ullah, S. E., Chupradit, S., & Minh Hieu, V., (2023). Impact of coal rents, transportation, electricity consumption, and economic globalization on ecological footprint in the USA. *Environmental Science and Pollution Research, 30*(15), 43040–43055.

60. Zhang, J., Mauzerall, D. L., Zhu, T., Liang, S., Ezzati, M., & Remais, J. V., (2010). Environmental health in China: progress towards clean air and safe water. *The Lancet, 375*(9720), 1110–1119.
61. Zhang, X., Yang, L., Li, Y., Li, H., Wang, W., & Ye, B., (2012). Impacts of lead/zinc mining and smelting on the environment and human health in China. *Environmental Monitoring and Assessment, 184*(1), 2261–2273.

INDEX

D

E

F

G

H

P

Q

R

S

T